The Exploit

Electronic Mediations

Katherine Hayles, Mark Poster, and Samuel Weber, Series Editors

continued on page 181

The Exploit

A Theory of Networks

Alexander R. Galloway and Eugene Thacker

Electronic Mediations, Volume 21

University of Minnesota Press
Minneapolis
London

Ideas in this book have been previously published in different form in the following essays cowritten by the authors: "Protocol and Counter-Protocol," in *Code: The Language of Our Time*, ed. Gerfried Stocker and Christine Schöpf (Linz: Ars Electronica, 2003); "Protocol, Control, and Networks," *Grey Room* 17 (Fall 2004); "In Defiance of Existence: Notes on Networks, Control, and Life-Forms," in *Feelings Are Always Local: DEAF04—Affective Turbulence*, ed. Joke Brouwer et al. (Rotterdam: V2_Publishing/NAi Publishers, 2004); "Networks, Control, and Life-Forms," in "Virtual Communities: Less of You, More of Us: The Political Economy of Power in Virtual Communities," ed. Jason Nolan and Jeremy Hunsinge, *SIGGROUP Bulletin* 25, no. 2 (2005), http://doi.acm.org/10.1145/1067721.1067722; "The Metaphysics of Networks," in *Censoring Culture: Contemporary Threats to Free Expression*, ed. Robert Atkins and Svetlana Mintcheva (New York: New Press, 2006), reprinted by permission of the New Press; "On Misanthropy," in *Curating Immateriality: The Work of the Curator in the Age of Network Systems*, ed. Joasia Krysa (New York: Autonomedia, 2006); "Language, Life, Code," *Architectural Digest*, September–October 2006.

Published by the University of Minnesota Press
111 Third Avenue South, Suite 290
Minneapolis, MN 55401-2520
http://www.upress.umn.edu

Library of Congress Cataloging-in-Publication Data

Galloway, Alexander R., 1974–
    The exploit : a theory of networks / Alexander R. Galloway and Eugene Thacker.
        p.   cm. — (Electronic mediations ; Vol. 21)
    Includes bibliographical references and index.
    ISBN: 978-0-8166-5043-9 (hc : alk. paper)
    ISBN: 978-0-8166-5044-6 (pb : alk. paper)

    1. Social networks.   2. Computer networks.   3. Computer network protocols.   4. Bioinformatics—Philosophy.   5. Sovereignty.
I. Thacker, Eugene.   II. Title.
    HM741.G34   2007
    303.48'3301—dc22   2007014964

Printed in the United States of America on acid-free paper

The University of Minnesota is an equal-opportunity educator and employer.

20 19 18 17 16 15 14   10 9 8 7 6 5 4

Contents

On Reading This Book

It is our intention in this book to avoid the limits of academic writing in favor of a more experimental, speculative approach. To that end, we adopt a two-tier format. Throughout Part I, "Nodes," you will find a number of condensed, italicized headers that are glued together with more standard prose. For quick immersion, we suggest skimming Part I by reading the italicized sections only. Alternatively, you may inspect the diversions and intensifications that form the main body of the text. Part II, "Edges," continues the experiment with a number of miniature essays, modules, and fragments. In this sense, we hope you will experience the book not as the step-by-step propositional evolution of a complete theory but as a series of marginal claims, disconnected in a living environment of many thoughts, distributed across as many pages.

Prolegomenon
"We're Tired of Trees"

In a recent e-mail exchange with the Dutch author and activist Geert Lovink, a person whose work we admire greatly, he made an interesting claim about the locus of contemporary organization and control. "Internet protocols are not ruling the world," Lovink pointed out, challenging our assumptions about the forces of organization and control immanent to a wide variety of networks, from biological networks to computer networks. Who is really running the world? "In the end, G. W. Bush is. Not Jon Postel," said Lovink, contrasting the American president with the longtime editor of the Internet network protocols.[1]

Lovink's claim that Internet protocols are not ruling the world strikes us as a very interesting thing to assert—and possibly quite accurate in many respects. The claim establishes one of the central debates of our time: the power relationship between sovereignty and networks. We interpret Lovink's claim like this: informatic networks are indeed important, but at the end of the day, sovereign powers matter more. The continual state of emergency today in the West, in the Middle East, in Africa, and in many other parts of the world is a testament to how much the various actions (and inactions) of sovereign powers indeed matter quite significantly. But is it really the case that networks matter less? And what kinds of networks—Postel's informatic networks, or the guerrilla networks of global terrorist groups? And what about sovereign powers who leverage the network form? Is the American government a network power? The United Nations?

Political discourse today generally slips into one of two positions: the first, often associated with the American state and its allies, maintains that "everything changed" with the fall of the Soviet Union, with the rise of the networked post-Fordist economies, and with September 11, 2001; the second, more associated with the critics of global empire, contends that the new millennial era is simply "more of the same."

To thumbnail our conversation with Lovink, one might associate him with the second position and us with the first. But this only reveals the complicated nature of the debate. America's neoconservative hawks often leverage the first position, using history's sharp cleavages as ammunition for more aggressive policies both foreign and domestic. And the reverse is true, too: it is the continuity of history—the "more of the same"—that fuels the American rhetoric of freedom, exported overseas just as it is exported to future generations. So on the one hand, there has been a great deal of attention through popular books, films, and television programs to the slipups and other controversies surrounding American foreign policy (in which everything from *Fahrenheit 911* to Abu Ghraib plays a part), heralding a growing crisis in Western sovereignty at the hands of various networked forces that seem to threaten it. But on the other hand, there is also a more cynical, business-as-usual stance, in which Western policies toward regions like the Middle East are seen as yet another, albeit crude, extension of American hegemony inaugurated decades ago in the aftermath of World War II.

One position tends to blame a particular administration for the state of things, while the other simply sees a repetition of a pattern that has been in place since World War II. While the first position tends to place excessive emphasis on a particular administration and leader, the second sees a progression, aided by Democrat and Republican alike, that has long been invested in the resources and political

opportunities that the Middle East affords. While the first position
says, "everything is different now," the second position says, "it's the
same as it ever was." Both positions concur on one point, however.
They agree on the *exceptional* character of the United States as the
sole global superpower in the wake of the Cold War.

*The rationale of "American exceptionalism" goes something like this:
the United States is an exception on the world stage; America's unique position as the world's only superpower gives it the prerogative not to comply
with multilateral institutions such as the International Court of Justice or
the Kyoto Protocol on climate change.*

In fact, in the period following September 11, 2001, the United
States argued that to maintain its own economic and political position, it had a certain *responsibility* to withdraw from multilateral treaties
that might open up the country to vindictive acts by lesser nations.
Compounding this political dynamic is the startling way in which the
United States has, throughout the last half century or so, dominated
the technology driving the world culture and economy, from the Windows operating system to Zoloft to the Boeing 747 aircraft. Thus the
idea of "American exceptionalism" is always refracted through two
crucial lenses of modernity: rapid technological change that, today at
least, centers around information networks, versus a continued expression of sovereignty alongside the emergence of these global networks.
And these two lenses are of course the same two positions we started
with: either everything is different, or nothing is different.

*Beyond international affairs, the theory of American exceptionalism
also has implications for a political theory of networks. In contrast to Lovink,
we maintain that in recent decades the processes of globalization have mutated
from a system of control housed in a relatively small number of power
hubs to a system of control infused into the material fabric of distributed
networks.*

This is illustrated in a number of examples: the decline of Fordist
economies in the West and the rise of postindustrial information and
service economies, the transnational and immigrant quality of labor

forces, the global outsourcing of production in high-tech fields, the dissemination of cultural products worldwide, the growing importance of networked machines in the military and law enforcement, the use of highly coded informatic systems in commodity logistics, or the deployment of complex pharmacological systems for health therapies and management of populations. Inside the dense web of distributed networks, it would appear that *everything is everywhere*—the consequence of a worldview that leaves little room between the poles of the global and the local. Biological viruses are transferred via airlines between Guangdong Province and Toronto in a matter of hours, and computer viruses are transferred via data lines from Seattle to Saigon in a matter of seconds. But more important, the solutions to these various maladies are also designed for and deployed over the same networks—online software updates to combat e-mail worms, and medical surveillance networks to combat emerging infectious diseases. The network, it appears, has emerged as a dominant form describing the nature of control today, as well as resistance to it.

But the U.S. empire and its America-first-and-only doctrine of sovereignty appear on the surface to contradict the foregoing picture of global informatic control.

American unilateralism seems to counter the notion that we live in a global network society. One might ask: How could there be a global system of distributed control if there also exists a single superpower? At the same time that there emerges a global world network, one also sees actions taken by the United States that seem to be the opposite of a network: the expression of a new sovereignty in the face of networks.

In this way, Lovink's initial claim—that the American president, not global networks, rules the world—sparks a whole series of questions for us. Is America a sovereign power or a networked power? Has sovereignty beaten back the once ascendant network form? Or has the network form invented a new form of sovereignty native to it?

First query: What is the profile of the current geopolitical struggle? Is it a question of sovereign states fighting nonstate actors? Is it a question of centralized armies fighting decentralized guerrillas? Hierarchies fighting networks? Or is a new global dynamic on the horizon?

Second query: Networks are important. But does the policy of American unilateralism provide a significant counterexample to the claim that power today is network based? Has a singular sovereignty won out in global affairs?

We cannot begin to answer these questions definitively. Instead we want to suggest that *the juncture between sovereignty and networks is the place where the apparent contradictions in which we live can best be understood.* It is the friction between the two that is interesting. Our choice should not simply be "everything is different" or "nothing has changed"; instead, one should use this dilemma as a problematic through which to explore many of the shifts in society and control over the last several decades.

Perhaps there is no greater lesson about networks than the lesson about control: networks, by their mere existence, are not liberating; they exercise novel forms of control that operate at a level that is anonymous and nonhuman, which is to say material.

The nonhuman quality of networks is precisely what makes them so difficult to grasp. They are, we suggest, a medium of contemporary power, and yet no single subject or group absolutely controls a network. Human subjects constitute and construct networks, but always in a highly distributed and unequal fashion. Human subjects thrive on network interaction (kin groups, clans, the social), yet the moments when the network logic takes over—in the mob or the swarm, in contagion or infection—are the moments that are the most disorienting, the most threatening to the integrity of the human ego. Hence a contradiction: the self-regulating and self-organizing qualities of emergent networked phenomena appear to engender and supplement the very thing that makes us human, yet one's ability to superimpose top-down control on that emergent structure evaporates in the blossoming of the network form, itself bent on eradicating the importance of any distinct or isolated node. This dissonance is most evident in network accidents or networks that appear to spiral out of control—Internet worms and disease epidemics, for instance. But calling such instances "accidents" or networks "out of control" is a misnomer. They are not networks that are somehow broken but *networks that work too well*. They are networks beyond one's capacity to control them, or even to comprehend them. At one moment the network appears far too large, as in the global dynamic of climate changes, but at another moment it appears too small, as with binary code or DNA. This is why we suggest that even while networks are entirely coincident with social life, networks also carry with them the most nonhuman and misanthropic tendencies. Indeed, sourcing the nonhuman within the human will be a major theme of this book.

So let us first outline a few provisional responses to the foregoing queries. While each is a useful cognitive exercise, we hope to show how each response is ultimately unsatisfying, and how a new approach is required for understanding the exceptional quality of sovereignty in the age of networks.

Provisional Response 1: Political Atomism (the Nietzschean Argument)

Action and reaction, force and counterforce—the argument can be made that the United States' decisions to declare war on terrorism or to intervene in the Middle East do not take place in a political vacuum. Perhaps the global machinations of "Empire" have elicited an American *ressentiment* in the form of unilateralism, a nostalgia for the good old days of the Cold War, when war meant the continued preparation for a standoff (never to arrive) between technologically advanced power blocs. Thus each advancement toward a decentralized global Empire consisting of France, Japan, Russia, and other leading industrialized nations is met by an American counterclaim to regain a singular world sovereignty.

However, this implies that, in contradistinction to the United States, the international community represented by the United Nations is the vanguard in the global political scene. The problem is that the very concept of a "united nations" is fraught with complication. On the one hand, there exists a romantic desire for a political tabula rasa, in which the many inequities between nations can be effaced by the "general will" of an international community. Yet on the other hand, the reality of how the United Nations operates is far from this: most nations are effectively excluded from the decision-making processes in international policy, and the stratifications between nations within the United Nations make for folly (either from within, as with criticism of U.S. war policy from Russia and China, or from without, as with the 2003 American veto of the nearly unanimous condemnation by UN member states of Israel's security fence in the occupied territories).

The United Nations is not simply the opposite of the United States, just as decentralized networks are not simply the opposite of centralized networks.

The Nietzschean argument, while it does call our attention to the political physics of action and reaction that exists within network structures, cannot account for conflict within networks, or better, between networks. Because its scope is so local, it can only account for the large-scale effects of network conflict by moving from local conflict to local conflict (in effect, moving from node to node). Nietzsche's notes in *The Will to Power* reveal this atomistic bias. Nietzsche begins from the analysis of "quanta of power" in constant interaction, and these quanta of power are understood somehow to compose the "will to power." Network structures challenge us to think about *what happens outside scale*—that is, between the jump from "quanta of power" to "will to power."

Provisional Response 2: Unilateralism versus Multilateralism (the Foucauldian Argument)

Critiques of U.S. unilateralism betray certain assumptions about power relationships. They tend to consider the output, or the "terminal effects," of power relations, rather than considering the contingencies that must be in place for those power effects to exist. For instance, the example of U.S. unilateralism is often used to demonstrate how power is not decentralized or network based, for, in this case, political power is encapsulated in one nation (or, more specifically, in the American president). The argument is similar to Lovink's claim at the beginning: yes, it is possible to acknowledge the networked character of power, but at the end of the day, it is the American president who makes decisions concerning war, resources, and trade. While this is undoubtedly true, it can also be argued that the political iconography of the presidency is the *effect*, not the cause, of asymmetrical global political relations. Furthermore, an effective political understanding of the situation cannot begin from the terminal effects of power relations. Instead one might ask: what power relationships need to be in place such that a single entity can obtain *propriety* over global organization and control?

While a Foucauldian emphasis on the bottom-up character of power relations is an important strategy for understanding the specifically global character of power relations, Foucault himself was always skeptical of contextualizing his work in terms of an ontology (and thus his emphasis on epistemology or on "power/knowledge").

> *A Foucauldian analysis may reveal how power is conditioned in its terminal effects (Homeland Security, the Patriot Act), yet it does not say much on the existence as such of this power.*

Put simply, such an analysis describes how power comes to be, but says little about *how it works* or even *that it exists* as such. A number of questions follow from this: What does it mean to "personify" or to individuate entities such as the United States in terms of unilateralism? (This is Michael Hardt and Antonio Negri's challenge as well, to argue that "America" and "Empire" are not mere synonyms.) More importantly, what are the networks of power relations that constitute the very ontology of the "unilateralism versus multilateralism" debate?

One immediate answer is given by the myriad political analyses concerning the United States.[2] These books hint at something intuitive: that the very term "American unilateralism" is a misnomer. Unilateralism works at several layers: on one layer it connects the White House with the House of Saud, on another with Israel, on another with Halliburton, on another with the United Nations, and so forth. So while it might sound like a contradiction, the Foucauldian analysis suggests that unilateralism must be understood as a network. This does not mean that it has no center; quite the opposite.

Following Foucault, to become unilateral, it is necessary to become multilateral, but via a veiled, cryptic sort of multilateralism. To become singular, one must become plural.

The center of so-called American unilateralism is constructed *through* its network properties. In a sense, any particular presidential administration is only half aware of this. It has placed itself in a paradoxical position. To ensure the cohesion of American unilateralism, it must forge links outside its domain.

Provisional Response 3: Ubiquity and Universality (the Determinist Argument)

Analyses of power relations often spend a great deal of time on the ideological content of political struggle: how the values of Islamic fundamentalism, or U.S. arrogance, produced within a certain historical context, lead to or justify violent actions. However, there is another view that focuses on the architecture of power, not just on its ideological content.

The role that communications and information networks have played in international terrorism and the "war on terror" has meant that media have now become a core component of war and political conflict.

One result of this view is that media can be seen to determine the very conditions of politics: the nexus of war and media makes war less real, while its effects, following Jean Baudrillard and Paul Virilio, take on the form of technological "accidents," "information bombs," and a "spirit" of terrorism of which the body is a vehicle. When the terms "power" and "control" are used in this context, this is really a shorthand for the material effects of media systems—the materiality of the media is, we are told, determinant of power relations, not the reverse. For instance, from this viewpoint the networks of FedEx or AT&T are arguably more important than that of the United States in terms of global economies, communication, and consumerism. This argument—what we might call a determinist argument—states that to understand the political situation, it is necessary to understand the material and technical infrastructures that provide the context for political conflict (we note this is not yet a causal argument). Consider debates around electronic wiretapping, data surveillance, human health protocols, biometrics, and information warfare. Suddenly the seemingly innocuous details of data packets, network protocols, and firewalls become politically charged indeed. Suddenly the comparison made at the outset between presidential power and technoscientific power makes a little more sense.

   The difficulty with the determinist approach (for the study of information technologies or for other areas) is that it often assumes a foundational position for the technology in question. Technology is assumed to, in effect, preexist politics.

   While we will not simply take the opposing position (that politics determine technology), there is something in this point of view worth noting, and these are the "ambient" or the "environmental" aspects of new media (however old they may be). New media are not just emergent; more importantly, they are everywhere—or at least that is part of their affect. Computers, databases, networks, and other digital technologies are seen to be foundational to contemporary notions of everything from cultural identity to war. Digital media seem to be everywhere, not only in the esoteric realms of computer animation, but in the *everydayness* of the digital (e-mail, mobile phones, the Internet). Within First World nations, this everydayness—this banality of the digital—is precisely what produces the effect of ubiquity, and of *universality*. (Is this a sufficient definition of network: that which is ubiquitous?) While the ubiquity of digital technology is undeniable, it is also far from being a global ubiquity, or from being the bedrock of society. We note a difference, then, between the ways in which new technologies can be *constitutive* of social, cultural, and political phenomena, and the notion that digital technologies are the *foundation* on which society is constructed. A determinist media studies approach cannot account for this difference, simply because it must, by definition, take for granted the ubiquity of its object of study. The implication is that technology *does* this or *does* that. However, we are not so sure that technology can be anthropomorphized so easily.

Provisional Response 4: Occultism and Cryptography (the Nominalist Argument)

A final interpretation should be noted. In all our discussion thus far, we have been assuming that the "United Nations" and the "United States" mean specific things. But in any contemporary situation, the issue of *naming,* or indeed the problem of substituting a title or name for a larger group, has to be recognized. After all, while it is the job of certain people to make decisions, the decisions made by "the American administration" are not decisions made solely by a single person or even a single entity. And yet our language shores up this assumption, in the very way that language personifies collective entities. This happens all the time. It becomes a sort of linguistic shorthand to say that the United States *does* this, or that al-Qaeda *does* that. Naming is nevertheless a tricky business; it leads to the problem of individuation. The United States' reaction to September 11, 2001, is indicative of the problem of individuation. Immediately after the attacks, two contradictory statements were insistently repeated in the media and by the government: that there is a "new enemy" constituting international networks of people, arms, money, information, and ideology; and that the name of this distributed new enemy is "terrorism" or even "al-Qaeda." In the same breath, we see the statement that our new enemy is networked and distributed to such a degree that *it cannot be named.* And yet there continues the persistent naming of the entity-that-cannot-be-named. What is obvious and immediate is the same thing that is shadowy and unknown. The plain is the obscure; the common is the cryptic.

But this is only the most explicit example of what is a much more basic issue concerning naming, especially when thinking about networks.

Any instance of naming always produces its shadowy double: nominalism, that is, the notion that universal descriptors do not adequately represent the referents they are supposed to name or demarcate.

Naming indicates its own impossibility. So perhaps the sovereign-network debate is really only a problem of naming. Perhaps it is a

nondebate. Is sovereignty nothing more than the *naming* of sovereignty? Derrida notes that part of the character of the name is its
refusal to be called by a name. Networks—be they terrorist networks
or networks of financial flows—further exacerbate this problem because of the simple fact that networks never claim to be integral
whole objects in the first place. To name a network is to acknowledge
a process of individuation ("the Internet," "al-Qaeda"), but it is also
to acknowledge the multiplicity that inheres within every network
("the Internet" as a meta-network of dissimilar subnets, "al-Qaeda"
as a rallying cry for many different splinter groups). This is why de-
veloping an *ontology* of networks—and not simply an ideology or a
technology of networks—is crucial to the current book and to our
understanding of the shape that global politics will take in the near
future. Everything's in a name. And everything's everything the name
is not. It is both referential (presupposing an already-existing thing
to which a name corresponds) as well as evocative (articulating a
foreground and a background where one did not previously exist).
The naming of the Internet in the popular imagination of the 1990s,
for instance, shows us the ontological effects that follow from the
name. Early so-called visionary writings about the Internet often chose
metaphors of interconnectivity to describe its potential, many of
them borrowed from neuroscience: the "World Brain," a "collective
intelligence," and so forth. Yet one denotes a very different sort of Internet when speaking about information security: computer "viruses,"
Internet "worms," and "computer immunology." These two examples
tell us already a number of things about the naming of networks,
least of which that naming often takes place in proximity to vital
forms (the Internet as both "brain" and "immunity"), and also that
naming often exists in relation to a particular social context (e.g., file
sharing or firewall intrusion). Is this the reason for our debate with
Lovink in which the only options are between one vital form (the
president "Bush") and another vital form (the network scientist "Postel")? Can a network ever be "faced" without also being named?

These represent a number of approaches to the queries mentioned at
the outset. But each in its own way falls short, especially when it

comes to thinking about networks as simultaneously technical and political. It will not do simply to assign a political content to a network form. Worse would be to claim that a network form is innately reactionary or progressive. It is foolish to fall back on the tired mantra of modern political movements, that distributed networks are liberating and centralized networks are oppressive. This truism of the Left may have been accurate in previous decades but must today be reconsidered.

To have a network, one needs a multiplicity of "nodes." Yet the mere existence of this multiplicity of nodes in no way implies an inherently democratic, ecumenical, or egalitarian order. Quite the opposite.

We repeat: the mere existence of networks does not imply democracy or equality. If anything, it is this existence-as-such of networks that needs to be thought; the existence of networks invites us to think in a manner that is appropriate to networks. (Would this then mean experimenting with something called "philosophy"?)

We hope that the present section offers a way to think topologically or "diagrammatically" about global political conflict.

"It is important to see today that the underlying technologies, the *network*, became a manifestation of an ideology itself," writes Pit Schultz. "Decentralization, or the rhizomatic swarm ideology is value free, useful for military, marketing, terrorism, activism and new forms of coercion. It is not equal to freedom, only in a mathematical sense."[3] Our approach here is above all a way of understanding certain historical realities through how they express the various attitudes, agents, resources, strategies, relationships, success criteria, and rules of engagement for specific types of struggle within the sociocultural sphere. By "thinking topologically," we mean an approach that compares the abstract spaces of different structural or architectonic systems. Pyramidal hierarchy and distributed networks, for example, have two different topologies of organization and control. The political dynamics of such topological analyses will form the backdrop of this book. "The novelty of the coming politics," writes Giorgio Agamben on the topological nature of political struggle, "is that it will no longer be a struggle for the conquest or control of the State, but a struggle between the State and the non-State (humanity)."[4]

While the new American exceptionalism is at the forefront of our thoughts today, we would like to situate it within a larger context by making reference to three diagrams for political conflict, each finding its own historical actualization: a politics of symmetry rooted in opposed power blocs, a politics of asymmetry in which power blocs struggle against insurgent networks, and a second model of symmetry in which networked powers struggle against other networked powers.

The first topology is a "politics of symmetry" perhaps best exemplified in the modern era by the symmetrical conflicts between the Soviets and the Americans, or earlier between the Allied and Axis powers. But this mode of political conflict was gradually superseded by a second one, in the second half of the twentieth century, which might be called the "politics of asymmetry." This is best exemplified in the guerrilla movements of the past several decades, or in terrorism, but also in the new social movements of the 1960s, and the newly networked societies of the 1990s. In all these examples, an asymmetrical conflict exists: grassroots networks posed against entrenched power centers. In an asymmetrical conflict, it is not possible to compare strategies of conflict one against the other. They are incommensurate. The conflict is actually rooted in asymmetry, without which there would be little antagonism. (It is not simply that feminism is opposed to patriarchy, but that they are asymmetrically opposed; racism and antiracism are not just opposed but exist in a relationship of asymmetry.) In conventional warfare, a networked insurgency will fail every time; however, in unconventional warfare (suicide bombing, hostage taking, hijacking, etc.), the insurgent is able to gain some amount of influence. Asymmetry—the diagram is the tactic. This second phase, roughly concurrent with what is called postmodernity, may best be understood through a proposition: postmodernity is characterized by frictions between structurally incommensurate political diagrams, where ultimately one is leveraged at the expense of the other (and something altogether different comes out the other end).

As stated previously, the high modern mode of political conflict is characterized by symmetrical war (power centers fighting power centers, Soviet and American blocs and so on). Then, in postmodernity, the latter decades of the twentieth century, one witnesses the rise of asymmetrical conflict (networks fighting power centers).[5] But after the postmodern mode of asymmetrical political conflict, and to bring the discussion up to the present day, we recognize in recent years the emergence of a new politics of symmetry. "What we are heading toward," write Hardt and Negri, "is a state of war in which network forces of imperial order face network enemies on all sides."[6]

This is why contemporary political dynamics are decidedly different from those in previous decades: there exists today a fearful new symmetry of networks fighting networks. One must understand how networks act politically, both as rogue swarms and as mainframe grids.

The network form has existed for quite some time, to be sure, but it has only recently attained any level of authority as a dominant diagram for mass social organization and control. In fact, the network form rose in power precisely as a corrective to the bloated union of hierarchy, decentralization, and bureaucracy that characterized the high modern period. For most of the second half of the twentieth century, the key dynamic revolved around new networks grappling with the old power hubs (the asymmetry described previously). But today a patchwork of new networked powers has emerged here and there and started to engage each other. In a sense, the power centers have evolved downward, adopting the strategies and structures of the terrorists and the guerrillas. The U.S. military has, for some time, shown an interest in and deployed modes of "infowar" and "cyberwar," and in the civilian sector we have seen new network power relations that span the spectrum technologically and ideologically: the increasing everydayness of surveillance (from Webcams to spyware to unmanned aerial vehicles [UAVs]), the socioeconomic push toward a mobile and wireless Internet, the cultural romanticism of flashmobs, and activist uses of mobile communications, hacking, and virtual sit-ins.

In this way, we should point out that the liberation rhetoric of distributed networks, a rhetoric famously articulated by Hans Magnus Enzensberger in his writings on the emancipation of media, is a foil for the real workings of power today. The rhetoric of liberation is also a foil for the real nature of threats. Whereas in former times, during the late modern era of the Cold War, for example, connectivity served to deaden the threat of weaponry and other forms of conflict—open up *channels* of communication with the Soviets, *engage* with China—today connectivity functions in exactly the opposite way.

Connectivity is a threat. The network is a weapons system.

The U.S. military classifies networks as weapons systems, mobilizing them as one would a tank or a missile. Today connectivity is a weapon. Bomb threats and terror alerts inject intangible anxiety into the population just as a real bomb might do. Media networks propagate messages from terrorists to all corners of the globe, just as airline networks propagate infectious diseases. The U.S. Department of Homeland Security is a reluctant proxy for al-Qaeda communiqués. Without connectivity, terrorism would not exist in its current form. It would be called something else—perhaps "revolt," "sedition," "murder," "treason," "assassination," or, as it was called during the period of the two world wars, "sabotage." Terrorism is quite at home in the age of distributed networks.

In this sense, the West created terrorism during the postmodern era, or at least created the conditions of possibility for terrorism to emerge.

Certainly the use of terror in ideological struggles predates postmodernity by decades if not centuries, and it is certainly not the West's conscious intent to bring terrorism into existence. So when we say the West invented terrorism, we mean this in the structural sense, not in the flimsy political sense of CIA "blowback," political resentment, or what have you. We mean that the West created terrorism in the same way that the overprescription of antibiotics creates new bacterial resistances. We mean that terrorism has evolved over time as a viable conflict strategy, one that is able to penetrate massified power blocs with extreme precision. As heterogeneity disappears, difference becomes all the more radical.

The more the West continues to perfect itself as a monolith of pure, smooth power, the greater the chance of a single asymmetrical attack penetrating straight to its heart.

The more Microsoft solidifies its global monopoly, the greater the chance for a single software exploit to bring down the entire grid. The more global health networks succeed in wiping out disease, the greater the chance for a single mutant strain to cause a pandemic. This is what we mean when we say the West created terrorism. The terrorist carries the day whether or not anyone dies. The stakes of the debate are forever changed.

The cruel truth is that terrorism works. But this is obvious, almost tautological, for in a networked milieu it cannot but work.

The power centers know this. And while the rhetoric of the American administration is about triumphing over the terrorists, the reality is that Homeland Security, the Pentagon, and many other state power structures are becoming more network oriented. "Assuming that Osama bin Laden's al-Qaeda network is our principal adversary, then we must outperform his network at all five levels at which information-age networks need to excel: the organizational, narrative, doctrinal, technological, and social," write the military strategists John Arquilla and David Ronfeldt about the West's current defensive posture. "Simply put, the West must build its own networks and learn to swarm the enemy network until it can be destroyed."[7]

When Arquilla and Ronfeldt warn that the West must "learn to swarm the enemy," they mean that the massified power blocs of the West must cease being massified power blocs. Centralized and decentralized architectures, which worked so well for so long during the modern period, are failing today, and thus the West must next

learn to succeed with a distributed architecture. The deployment of
the most highly standardized and controlled technology in history, the
Internet, at the end of the twentieth century is but a footnote to this
general historical transformation. Arquilla and Ronfeldt urge the
West to become network oriented—flexible, distributed, agile, robust, disseminated, invisible—in order to grapple on equal footing
with the cellular, distributed, networked architectures of the terrorists. Again, it is a question of networks fighting networks. Massified
power blocs are unable to beat networks, the authors argue.

*Once this new network–network symmetry is firmly ensconced in one's
imagination, it should just as quickly be qualified, for one of the crucial rules
of networks is that they are internally variable. Not all networks are equal.
(Thus, from a certain perspective, networks are internally asymmetrical.)*

Networks are never consistent or smooth but exhibit power relations that are internally inconsistent. This is what *makes* them networks. The new network–network symmetry does not mean homogeneity. (The older, bilateral conflict—of the Cold War era, for
example—is by definition nonnetworked because it is symmetrical
and consistent.) Networked power has learned from history and may
use all varieties of authority and organization: centralized, decentralized, distributed, violent, coercive, desiring, liberating, and so on.

*Networked power is additive, not exclusive. It propagates through "and,"
not "or."*

In being additive, networks necessarily establish unequal power
differentials within their very structures. Just as the mere existence of
the Internet doesn't imply democracy, networks don't imply the distributive mode. Nor do networks imply total horizontality. The point
is that not all networks are the same. Clay Shirky has written on how a
power law distribution of resources naturally springs up in distributed systems: "In systems where many people are free to choose between many options, a small subset of the whole will get a dispropor-
tionate amount of traffic (or attention, or income), even if no
members of the system actively work toward such an outcome. This

has nothing to do with moral weakness, selling out, or any other psychological explanation. The very act of choosing, spread widely enough and freely enough, creates a power law distribution."[8] Power law distributions help explain America's place in the global network. Even in distributed networks, certain power centers will necessarily emerge through a sort of clustering pattern, just as certain Web sites will emerge as supernodes within the larger net. Today these power centers are called Bangalore, or Microsoft, or Archer Daniels Midland. But the fact that they have names does not discount the continuing affective force of the distributed networks they inhabit and move through.

Since networked power is additive in its political strategies, control in the information age is created through the selective articulation of certain tactics here and others there.

It is a case not of distributive control winning out over centralized sovereignty but of the orchestrated use of one against the other. In this sense, it is necessary for networks to exist in order that sovereignty may be created. In the rich philosophical literature on political sovereignty, sovereignty is always compromised by "the outside." In Aquinas the earthly sovereign must still answer to God. In the secularized version of Bodin, sovereignty is limited by the good of the commonwealth (as is the case for Spinoza). In Grotius sovereignty of one nation is always tempered by the sovereignty of other nations and the possibility of war.

Today this same dynamic is at play. Networked power is based on a dialectic between two opposing tendencies: one radically distributes control into autonomous locales; the other focuses control into rigidly defined hierarchies. All political regimes today stand in some relation to networks. So it is possible to have unilateralism and networks, a fact that makes the American regime so beguiling.

When a sovereign networked power is able to command globally and instantaneously, there exist what might be called "global-single" command events. By "global-single" we mean that while the networked

sovereign is globally networked, extending into all countries and all
social contexts as a localized native (not as an interloper or occupying force), it is at the same time able to emit a single command decision, from a single location. The network is smart enough to distinguish between distributed organizational mechanics and directed
commands, for in fact they are the same thing. In this way, a networked
sovereign is able to "flip the switch" and watch as command decisions
propagate rapidly around the globe. An example today would be something like an operating system security update. It emanates from a
single networked sovereign (a software company) but is a type of
material command that is issued in a global, relatively instantaneous
manner. Of course, the antagonists in this scenario, e-mail worms or
zombie botnets, operate through exactly the same logic. Global-single
commands require an exceedingly high level of technological organization and interconnection. Today they appear mainly as exploita-
tions of technoscientific "opportunities" or are issued as patches or
therapies to combat such material threats, but they may also be issued
in the cultural sectors, representing a greater evolution of today's
strategies of "just-in-time production" and trend spotting and rapid
response. If networked sovereigns wield tactics like the global-single
command event, it is interesting to ponder what an asymmetrical response to such an event would look like. Geography alone will not
save you. Nor will agility or guile. These are the questions addressed
in the following sections on liberation and counterprotocological
practices.

*So networks and sovereignty are not incompatible. In fact, quite the
opposite: networks create the conditions of existence for a new mode of
sovereignty. America is merely the contemporary figurehead of sovereignty-
in-networks.*

While in the past networks may have posed significant threats to
power in everything from the grassroots social movements of the 1960s
to guerrilla armies and terrorist organizations, it is not the case today.
Networks are not a threat to American power. In fact, the opposite is
true: networks are the medium through which America derives its

sovereignty. In the 1970s, when Gilles Deleuze and Félix Guattari wrote about "the war machine" as a resistance to the state apparatus, they were describing the threat that an elusive network can pose to a power center. But by 1990, Deleuze had recognized the historical transformation that had taken place in the intervening years, and he wrote of the new network form in terms of "ultrarapid forms of apparently free-floating control."[9]

What the United States accomplished in the years after 1989 was to derive its own sovereignty from within the "ultrarapid" and "free-floating" networks. This results in the curious dual rhetoric of the "international presence" in peacekeeping operations combined with an "American-led" force, an equivocation held together only by the most flimsy political fantasy. This flimsy assimilation is precisely the model for sovereignty in networks.

The current American regime is in the political vanguard. It aims to establish sovereignty in a new political structure that is antithetical to traditional modes of sovereignty.

The trick is to reach beyond a theory of "power law distributions" to an actual theory of political action rooted in networks. Now we can return to our original constellation of queries: What is the nature of the current geopolitical struggle? Is the United States an exception on the world stage? Has a singular sovereignty returned to networks, global affairs? Our argument has three steps:

1. The modern period is characterized by both symmetrical political conflicts waged by centralized power blocs, and also asymmetrical political conflicts in which networked actors struggle against centralized powers.

Many have further suggested that asymmetric conflict is in fact a historical response to the centralization of power. This type of asymmetric intervention, a political form bred into existence as the negative likeness of its antagonist, is the inspiration for the concept of "the exploit," a resonant flaw designed to resist, threaten, and ultimately desert the dominant political diagram. Examples include the suicide bomber (versus the police), peer-to-peer protocols (versus the music conglomerates), guerrillas (versus the army), netwar (versus cyberwar), subcultures (versus the family), and so on.

   2. The present day is symmetrical again, but this time in the symmetrical form of networks fighting networks.

A new sovereignty, native to global networks, has recently been established, resulting in a new alliance between "control" and "emergence." Networks exist in a new kind of global universalism, but one coextensive with a permanent state of internal inconsistency and exceptionalism. In this network exception, what is at stake is a newly defined sense of nodes and edges, dots and lines, things and events—networked phenomena that are at once biological and informatic.

   3. To be effective, future political movements must discover a new exploit.

A wholly new topology of resistance must be invented that is as asymmetrical in relationship to networks as the network was in relationship to power centers. Resistance *is* asymmetry. The new exploit will be an "antiweb." It will be what we call later an "exceptional topology." It will have to consider the radically *unhuman* elements of all networks. It will have to consider the nonhuman within the human, the level of "bits and atoms" that are even today leveraged as value-laden biomedia for proprietary interests. It has yet to be invented, but the newly ascendant network sovereigns will likely breed the antiweb into existence, just as the old twentieth-century powers bred their own demise, their own desertion.

PART I

Nodes

The quest for "universals of communication" ought to make us shudder.

—*Gilles Deleuze*

Nodes

*Discourse surrounding networks, in keeping with the idea of networks them-
selves, is becoming more and more ubiquitous.*

For the last decade or more, network discourse has proliferated
with a kind of epidemic intensity: peer-to-peer file-sharing networks,
wireless community networks, terrorist networks, contagion networks
of biowarfare agents, political swarming and mass demonstration,
economic and finance networks, online role-playing games, personal
area networks, mobile phones, "generation Txt," and on and on.

*Often the discourse surrounding networks tends to pose itself both
morally and architecturally against what it sees as retrograde structures
like hierarchy and verticality.*

These structures are seen to have their own concomitant techniques for keeping things under control: bureaucracy, the chain of
command, and so on. "We're tired of trees," wrote Deleuze and Guattari. But even beyond the fields of technology and philosophy, the
concept of the network has infected broad swaths of contemporary

life. Even the U.S. military, a bastion of vertical, pyramidal hierarchy, is redefining its internal structure around network architectures, as the military strategists Arquilla and Ronfeldt have indicated in their work. They describe here a contemporary mode of conflict known as "netwar": "Netwar is about the Zapatistas more than the Fidelistas, Hamas more than the Palestine Liberation Organization (PLO), the American Christian Patriot movement more than the Ku Klux Klan, and the Asian Triads more than the Cosa Nostra."[1] These in/out lists are, of course, more fun to read than they are accurate political evaluations, but it is clear that the concept of connectivity is highly privileged in today's societies.

In fact, the idea of connectivity is so highly privileged today that it is becoming more and more difficult to locate places or objects that don't, in some way, fit into a networking rubric.

This is particularly the case as the Fidelistas and so on are further eclipsed by their network-savvy progeny. The 2001 USA PATRIOT Act and other legislation allowing increased electronic surveillance further reinforce the deep penetration of networked technologies and networked thinking. One wonders if, as networks continue to propagate, there will remain any sense of an "outside," a nonconnected locale from which we may view this phenomenon and ponder it critically.

In today's conventional wisdom, everything can be subsumed under a warm security blanket of interconnectivity. But this same wisdom hasn't yet indicated quite what that means, nor how one might be able to draft a critique of networks.

All this fanfare around networks highlights the continued indissociability of politics and technology. There are several sides to the debate. The technophilic perspective, such as that expressed by Howard Rheingold or Kevin Kelly, is an expression of both a technological determinism and a view of technology as an enabling tool for the elevation of bourgeois humanism in a broadly general sense. The juridical/governance perspective, seen in the work of Lawrence Lessig, Yochai

Benkler, and others, posits a similar situation whereby networks will bring about a more just and freer social reality via legal safeguards. The network science perspective, expressed in popular books by Mark Buchanan or Albert-László Barabási, portrays the network as a kind of apolitical natural law, operating universally across heterogeneous systems, be they terrorism, AIDS, or the Internet. Moreover, this dichotomy (between networks as political and networks as technical) is equally evident in a variety of other media, including news reportage, defense and military research, and the information technology industries.

Yet this "network fever" has a tendency to addle the brain, for we identify in the current literature a general willingness to ignore politics by masking them inside the so-called black box of technology.[2]

Thus one of our goals is to provide ways of critically analyzing and engaging with the "black box" of networks, and with this ambivalence between politics and technology (in which, sadly, technology always seems to prevail).

The question we aim to explore here is: what is the principle of political organization or control that stitches a network together?

Writers like Michael Hardt and Antonio Negri have helped answer this question in the sociopolitical sphere. Their concept of "empire" describes a global principle of political organization. Like a network, empire is not reducible to any single state power, nor does it follow an architecture of pyramidal hierarchy. Empire is fluid, flexible, dynamic, and far-reaching. In that sense, the concept of empire helps us greatly to begin thinking about political organization in networks.

But are networks always exclusively "human"? Are networks misanthropic? Is there a "nonhuman" or an "unhuman" understanding of networks that would challenge us to rethink the theory and practice of networks?

While we are inspired by Hardt and Negri's contribution to political philosophy, we are concerned that no one has yet adequately answered this question for the technological sphere of bits and atoms. That is, we seek a means of comprehending networks as simultaneously material and immaterial, as simultaneously technical and political, as simultaneously misanthropic and all-too-human.

Let us continue then not with an empirical observation but with a concept. Derived from the discourses of both the life sciences and computer science, the concept of "protocol" refers to all the technoscientific rules and standards that govern relationships within networks. Protocols abound in technoculture. They are rooted in the laws of nature, yet they sculpt the spheres of the social and the cultural. They are principles of networked interrelationality, yet they are also principles of political organization.

Quite often networked relationships come in the form of communication between two or more computers, but the relationships can also refer to purely biological processes, as in the systemic phenomenon of gene expression or the logics of infection and contagion. Protocol is not a single thing but a set of tendencies grounded in the physical tendencies of networked systems. So by "networks" we mean any system of interrelationality, whether biological or informatic, organic or inorganic, technical or natural—with the ultimate goal of undoing the polar restrictiveness of these pairings.

Abstracted into a concept, protocol may be defined as a horizontal, distributed control apparatus that guides both the technical and political formation of computer networks, biological systems, and other media.

Molecular biotechnology research frequently uses protocols to configure biological life as a network phenomenon, whether in gene expression networks, metabolic networks, or the circuitry of cell signaling pathways. In such instances, the biological and the informatic become increasingly enmeshed in hybrid systems that are more than biological: proprietary genome databases, DNA chips for medical diagnostics, and real-time detection systems for biowarfare agents. Likewise in computer networks, science professionals have, over the years, drafted hundreds of protocols to create e-mail, Web pages, and so on, plus many other standards for technologies rarely seen by human eyes. An example might be the "Request for Comments" series of Internet white papers, the first of which was written by Steve Crocker in 1969, titled "Host Software."[3] Internet users commonly use protocols such as http, FTP, and TCP/IP, even if they know little about how such technical standards function. If networks are the structures that connect organisms and machines, then protocols are the rules that make sure the connections actually work.

Protocol is twofold; it is both an apparatus that facilitates networks and a logic that governs how things are done within that apparatus.

Today network science often conjures up the themes of anarchy, rhizomatics, distribution, and antiauthority to explain interconnected systems of all kinds. Our task here is not to succumb to the fantasy that any of these descriptors is a synonym for the apolitical or the disorganized, but in fact to suggest the opposite, that rhizomatics and distribution signal a new management style, a new physics of organization that is as real as pyramidal hierarchy, corporate bureaucracy, representative democracy, sovereign fiat, or any other principle of social and political control. From the sometimes radical prognostications of the network scientists, and the larger technological discourse of thousands of white papers, memos, and manuals surrounding it, we can derive some of the basic qualities of the apparatus of organization that we here call protocol:[4]

- Protocols emerge through the complex relationships between autonomous, interconnected agents.
- To function smoothly, protocological networks must be robust and flexible; they must accommodate a high degree of contingency through interoperable and heterogeneous material interfaces.
- Protocological networks are inclusive rather than exclusive; discrimination, regulation, and segregation of agents happen on the inside of protocological systems (not by the selective extension or rejection of network membership to those agents).
- Protocols are universal and total, but the diachronic emergence of protocols is always achieved through principles of political liberalism such as negotiation, public vetting, and openness.

- Protocol is the emergent property of organization and control in networks that are radically horizontal and distributed.

Each of these characteristics alone is enough to distinguish protocol
from many previous modes of social and technical organization (such
as hierarchy or bureaucracy). Together they compose a new, sophisticated system of distributed control. As a technology, protocol is im-
plemented broadly and is thus not reducible simply to the domain of
institutional, governmental, or corporate power.

*In the broadest sense, protocol is a technology that regulates flow, directs
netspace, codes relationships, and connects life-forms.*

Networks always have several protocols operating in the same
place at the same time. In this sense, networks are always slightly
schizophrenic, doing one thing in one place and the opposite in another. The concept of protocol does not, therefore, describe one all-
encompassing network of power—there is not one Internet but many
internets, all of which bear a specific relation to the infrastructural
history of the military, telecommunications, and science industries.
This is not a conspiracy theory, nor is it a personification of power.
Protocol has less to do with individually empowered human subjects
(the pop-cultural myth of hackers bringing down "the system") who
might be the engines of a teleological vision for protocol than with
manifold modes of individuation that arrange and remix both human
and nonhuman elements (rather than "individuals" in the liberal
humanist sense). But the inclusion of opposition within the very fabric
of protocol is not simply for the sake of pluralism—which of course
it leverages ideologically—but instead is about politics.

*Protocological control challenges us to rethink critical and political ac-
tion around a newer framework, that of multiagent, individuated nodes in
a metastable network.*

Political action in the network, then, can be guided deliberately
by human actors, or accidentally affected by nonhuman actors (a
computer virus or emerging infectious disease, for example). Often,
tactical misuse of a protocol, be it intended or unintended, can iden-

tify the political fissures in a network. We will suggest later that such moments, while sometimes politically ambiguous when taken out of context, can also serve as instances for a more critical, more politically engaged "counterprotocol" practice. As we shall see, protocological control brings into existence a certain contradiction, at once distributing agencies in a complex manner while at the same time concentrating rigid forms of management and control. This means that protocol is less about power (confinement, discipline, normativity), and more about control (modulation, distribution, flexibility).

Technology (or Theory)

There exists an entire science behind networks, commonly known as graph theory, which we would like to briefly outline here, for it subtends all our thinking on the nature of networks and systems.[5] Mathematically speaking, a graph is a finite set of points connected by a finite set of lines. The points are called "nodes" or vertices, and the lines are called "edges." For the sake of convenience we will use G to refer to a graph, N to refer to the nodes in the graph, and E to refer to its edges. Thus a simple graph with four nodes (say, a square) can be represented as $N = \{n_1, n_2, n_3, n_4\}$ and its edges as $E = \{(n_1, n_2), (n_2, n_3), (n_3, n_4), (n_4, n_1)\}$. In a graph, the number of nodes is called the "order" (in the square example, $|N| = 4$), and the number of edges is called the "size" ($|E| = 4$).

In the mathematical language of graph theory, networks provide us with a standard connect-the-dots situation.

Given this basic setup of nodes and edges, a number of relationships can be quantitatively analyzed. For instance, the "degree" of a node is the number of edges that are connected to it. A "centralized" or "decentralized" graph exists when a relatively small number of nodes function as "hubs" by having many edges connected to them, and when the remaining "leaf" nodes have only one edge. This results in a graph where the order and size are roughly the same. Likewise, a "distributed" graph exists when the hub/leaf split disappears and all nodes have approximately the same degree. This results in a graph where the size far exceeds the order. What can we tell by both the order and size of a graph? One of the basic theorems of graph theory states that for any graph with a finite number of edges, the sum of the degrees of the nodes equals twice the number of edges. That is, if the degree of any node is the number of edges connected to it (for node n_1 with two edges connected to it, its degree = 2), the sum of all the degrees of the graph will be double the size of the graph (the number of edges). For a square, the sum of the degrees is 8 (the nodes [the square's corners] each have 2 edges [the square's lines] connected to them), while the sum of the edges is 4. In other words, the *connectivity* of a graph or network is a value different from a mere count of the number of edges. A graph not only has edges between nodes but also has edges connecting nodes.

From a graph theory perspective, networks can be said to display three basic characteristics: their organization into nodes and edges (dots and lines), their connectivity, and their topology. The same sets of entities can result in a centralized, rigidly organized network or in a distributed, highly flexible network.

The institutional, economic, and technical development of the Internet is an instructive case in point. While the implementation of packet-switching technology in the U.S. Department of Defense's ARPANET ostensibly served the aims of military research and security, that network also developed as a substantial economic network, as well. Paul Baran, one of the developers of packet switching, uses basic graph theory principles to show how, given the same set of nodes or points, and a different set of edges or lines, one gets three very different network topologies.[6] The familiar distinction between centralized, decentralized, and distributed networks can be found everywhere today, not only within computer and information technologies but also in social, political, economic, and biological networks.

As we have suggested, networks come in all shapes and flavors, but common types include centralized networks (pyramidal, hierarchical schemes), decentralized networks (a core "backbone" of hubs each with radiating peripheries), and distributed networks (a collection of node-to-node relations with no backbone or center).

From the perspective of graph theory, we can provisionally describe networks as metastable sets of variable relationships in multinode, multiedge configurations.

In the abstract, networks can be composed of almost anything: computers (Internet), cars (traffic), people (communities), animals (food chains), stocks (capital), statements (institutions), cultures (diasporas), and so on. Indeed, much of the research in complex dynamic systems, nonlinear dynamics, and network science stresses this convergence of heterogeneous phenomena under universal mathematical principles.

However, we stress this point: graph theory is not enough for an understanding of networks; or rather, it is only a beginning.

Although graph theory provides the mathematical and technical underpinning of many technological networks (and the tools for analyzing networks), the assumptions of graph theory are equally instructive for what they omit.

First is the question of agency. The division between nodes and edges implies that while nodes refer to objects, locations, or space, the definition of edges refers to actions effected by nodes. While agency is attributed to the active nodes, the carrying out of actions is attributed to the passive edges (the effect of the causality implied in the nodes). Graphs or networks are then diagrams of force relationships (edges) effected by discrete agencies (nodes). In this, graphs imply a privileging of spatial orientations, quantitative abstraction, and a clear division between actor and action.

Second is what might be called the "diachronic blindness" of graph theory. Paradoxically, the geometrical basis (or bias) of the division between "nodes" and "edges" actually works against an understanding of networks as sets of relations existing in time. While a graph may evoke qualities of transformation or movement in, for example, the use of directed edges, it is an approach that focuses on fixed "snapshot" modeling of networked ecologies and their simulation using mathematical models and systems. This is, we suggest, a fundamentally synchronic approach.

Related to this is the pervasive assumption that networks can exist in an ideal or abstract formulation (a mathematical graph) estranged from the material technologies that, in our view, must always constitute and subtend any network.

A final disadvantage of graph theory is the question of internal complexity and topological incompatibility. Not only are networks distinguished by their overall topologies, but networks always contain several coexistent, and sometimes incompatible, topologies. This is a lesson learned from general systems theory, whereby networks consist of aggregate interconnections of dissimilar subnetworks. The subnet topologies will often be in transition or even be in direct opposition to other forms within the network. Thus any type of protocological control exists not because the network is smooth and continuous but precisely because the network contains within it antagonistic clusterings, divergent subtopologies, rogue nodes. (This is what makes them networks; if they were not internally heterogeneous, they would be known as integral wholes.) For example, a merely "technical" description of the topology of the Internet might describe it as distributed (for example, in the case of peer-to-peer file-sharing networks based on the Gnutella model, or in the routing technologies of the Internet protocol). But it is impossible to disassociate this technical topology from its motive, use, and regulation, which also make it a social topology of a different form (file-sharing communities), an economic topology with a still different form (distribution of commodities), and even a legal one (digital copyright). All of these networks coexist, and sometimes conflict with each other, as the controversy surrounding file sharing has shown. While graph theory can indeed model a number of different topologies, we prefer an approach wherein the coexistence of multiple incompatible political structures is assumed as fundamental.

Thus not only do existing network theories exclude the element that makes a network a network (its dynamic quality), but they also require that networks exist in relation to fixed, abstract configurations or patterns (either centralized or decentralized, either technical or political), and to specific anthropomorphic actors.

Indeed, one of the arguments presented here is to reinforce the notion that material instantiation is coextensive with pattern formation. Material substrate and pattern formation exist in a mutually reciprocal relationship, a relationship that itself brings in social-political and technoscientific forces.

Theory (or Technology)

In the "Postscript on Control Societies," a delectably short essay from 1990, Deleuze defines two historical periods: first, the "disciplinary societies" of modernity, growing out of the rule of the sovereign, into the "vast spaces of enclosure," the social castings and bodily molds that Michel Foucault has described so well; and second, what Deleuze terms the "societies of control" that inhabit the late twentieth century—these are based around protocols, logics of "modulation," and the "ultrarapid forms of free-floating control."[7] For Deleuze, "control" means something quite different from its colloquial usage (as in "control room" or "remote control").

Control is not simply manipulation, but rather modulation.

One does not simply control a device, a situation, or a group of people; rather, "control" is what enables a relation to a device, a situation, or a group. "People are lines," Deleuze suggests. As lines, people thread together social, political, and cultural elements. While in disciplinary societies individuals move in a discrete fashion from one institutional enclosure to another (home, school, work, etc.), in the societies of control, individuals move in a continuous fashion between sites (work-from-home, distance learning, etc.). In the disciplinary societies, one is always starting over (initiation and graduation, hiring and retirement). In the control societies, one is never finished (continuing education, midcareer changes). While the disciplinary societies are characterized by physical semiotic constructs such as the signature and the document, the societies of control are characterized by more immaterial ones such as the password and the computer.[8]

The problem of "control" in networks is always doubled by two perspectives: one from within the network and one from without the network. Networks are, in this sense, the horizon of control.

On the one hand, control is tantamount to forms of network management, for control in networks must meet the challenge of network regulation from a site that is internal to the network—the most "controlled" control would be one that pervades the network itself. Control in networks must aim for an effectiveness that is immanent to the network, in the sense that the most perfectly controlled network is one that controls or regulates itself. But, on the other hand, control in networks is always counterbalanced by another challenge: to be effective from outside the network (either as a set of meta-guidelines or as being logically "above" the network itself). The network itself must be articulated as an object of design, implementation, and regulation. Control in this sense does not pervade the network but operates over it; control in this sense is topsight and oversight.

The breakdown of disciplinary societies and the emergence of control societies raise a whole host of philosophical problems, problems that are both absolutely "ancient" and contemporary. Take, for example, the notion of "substance."

Classical philosophers from the pre-Socratics to Aristotle mused a great deal on substance. They asked: Of what is the world made? What is the fundamental property of, for example, a living creature that allows us to conceive of and say "creature"? The question of substance is not a question of being: it is not *that* it exists but rather *how* it exists.[9] The question is not "what is it?" but rather "how does it work?"

The question of substance poses particular problems when thinking about networks. Is it safe to define a network as a substance, as a particular thing? We can ask: Of what is a network made? Is it enough to say that a network is made of fiber-optic cable, routers, and terminals? This would limit our concept of "network" to computer networks. Would it be enough to expand this to include organisms, cells, and proteins? Is it thus the more abstract notion of "nodes" and "edges" we noted earlier? This alone would be too general, for potentially anything and everything could be conceived of as a node or an edge (if everything is a network, then nothing is).

Should we define an essential property—"relation" or "interrelation"—and construct a concept of the network from that? This could provide a starting point, but defining essences is always a tricky business. Relation always presupposes at least two "things" that are related. Relation is not, then, a "thing" but the relation between things. Is it a gap, an interval, a synapse? We are led into even more treacherous waters: relation is "the nothing" between two things. Following such a line of argument, our notion of "network" would be founded on the most insubstantial of substances.

Like the concept of substance, the problem of "individuation" is also a long-standing concept in philosophical thought (we will return to this later in a different vein). And, like substance, individuation is a concept that is equally filled with aporias. But unlike substance, these aporias are generative and evocative rather than reifying or reductive.

To individuate is to posit both the specific and the generic.

For instance, if one says, "I am reading this book," the "book" in the statement is implicitly one of a general category of objects called books, as well as an explicit reference to a specific and singular book (not just any book, but this one here in hand). Late medieval philosophy, influenced greatly by Aristotle's *Categories*, debated individuation at length. At the most general level, individuation is about what makes a thing what it is: what is it that makes a "cloud" a cloud? More specifically, individuation also has to do with predication: what is it that makes "Socrates" a man? But individuation is not simply about language (subject and predicate), for it brings together the concept (the concept of clouds), the thing (a specific cloud, that one up there), and language ("cloud") into an isomorphic field that bypasses later philosophical debates about language and the "thing in itself."[10]

> *A mode of individuation may produce a distinct person, a mass of people, a nation-state, a corporation, a set of gadgets, animals, plants, or any formation of matter.*

Subjects as individual people, then, are particular modes of individuation to which sets of values are ascribed: agency, autonomy, self-consciousness, reason, emotion, rights, and so on. Although "individuation" is a well-worn philosophical concept, in the context of the control societies, individuation is assumed to be continually modulated, precisely because it is informatic, statistical, and probabilistic.

> *Perhaps it is best to define a network as a mode of individuation? But if so, how is a network individuated? What makes a network "a" network? What is the "circumference" of a network?*

These almost geometrical quandaries become even more relevant when couched in the language of political philosophy: What is inside a network? What is outside? This is not simply a question about who gets access to a network or about who decides what to include or exclude from a network. Such an approach presumes the prior existence of a network, and then, only after this, is access or inclusion raised as a problem. Instead, the question of individuating a network is really a problem of establishing the very conditions in which a network can exist at all. It is, in other words, a problem of sovereignty.

Traditional concepts of sovereignty are often juridical in nature—that is, they define sovereignty as the ability to exercise control over bodies and resources based on law, or, as Foucault put it, the authority to "take life or let live." By contrast, contemporary political thought often defines sovereignty not as the power to command or execute a law but as the power to claim exceptions to the rule.[11] The sovereign ruler occupies a paradoxical position, at once within the law (in that the ruler forms part of the body politic), and yet outside the law (in that the sovereign can decide when the law no longer applies). Sovereignty is, then, not power or force but the ability to decide—in particular, the ability to decide what constitutes an exceptional situation, one that calls for a "state of exception" and a suspension of the law.[12] But it is not always clear where the line between "exception" and "rule" lies. The notion of a "permanent state of emergency" is one consequence of this definition of sovereign power. If this is the case, then a central challenge for any radical politics today is explaining the strange intimacy between the sovereign "state of exception" and the decentralized character of global networks for which "exceptionalism" is formally necessary.

The tension we noted within "control" —at once inside and outside the network, at once "within" and "above" —can be rephrased as a question about sovereignty. The quandary is this: no one controls networks, but networks are controlled.

And we stress that no "one" controls networks because they deindividuate as much as they individuate. Networks individuate within themselves (stratifying different types of nodes, different types of users, different types of social actors), and they also auto-individuate as well (the systems of "small worlds" or "friends of friends" described in social network theory). But these processes of individuation are always accompanied by processes of deindividuation, for each individuation is always encompassed by the "mass" and aggregate quality of networks as a whole, everything broken down into stable, generic nodes and discrete, quantifiable edges. Nodes are erased as quickly as edges are established, hierarchies exist within networks, "horizontal" decentralization interferes with "vertical" centralization, topologies become topographies . . .

In the control society, what is the difference between sovereignty and control? That is, does sovereignty exist in networks?

If we are to take seriously the networked view of power relations, then individuals would need to be considered not as individuals but as what Deleuze calls "dividuals": "In control societies . . . the key thing is no longer a signature or number but a code: codes are *passwords*, whereas disciplinary societies are ruled (when it comes to integration by resistance) by *precepts*. The digital language of control is made of codes indicating where access to some information should be allowed or denied. We're no longer dealing with a duality of mass and individual" from the modern era. Instead "individuals become '*divid-uals*,' and masses become samples, data, markets, or '*banks*.'"[13]

What follows from this is that control in networks operates less through the exception of individuals, groups, or institutions and more through the exceptional quality of networks or of their topologies. What matters, then, is less the character of the individual nodes than the topological space within which and through which they operate as nodes. To be a node is not solely a causal affair; it is not to "do" this or "do" that. To be a node is to exist inseparably from a set of possibilities and parameters—to function within a topology of control.

Not all topologies are equal; some are quite exceptional, existing for short periods of time (e.g., a highly centralized organization may briefly become decentralized to move its operations or internally re-structure). But every network has its own *exceptional topology*, the mode of organization that is uncommon to itself. Distributed networks, be they computer based or community based, must at some point confront the issue of "decision," even if the decision is to become a network itself. If the network is anthropomorphized, such decision points require centralization, a single point from which the decision can be made. (Sometimes this is called "the central nervous system" or "the standards-setting community.") The point at which sover-eignty touches network control may very well lie in this notion of an exceptional network, an exceptional topology. In the case of con-temporary politics, America's networked power rises only in direct proportion to the elimination, exclusion, and prohibition of net-worked power in the guerrilla and terrorist movements.

Perhaps we are witnessing a sovereignty that is unlike the traditional forms of sovereignty, a mode of sovereignty based not on exceptional events but on exceptional topologies.

Without a doubt, these exceptional topologies are troubling. They exercise sovereignty, and yet there is no one at the helm making each decision. One might call these societies "misanthropic" or "anti-anthropological." The societies of control have an uncanny ability to elevate nonorganic life, placing it on par with organic life. And yet there is a sense in which networks remain dynamic, always changing,

modulating, in flux, *alive*. If the body in disciplinary societies is pre-
dominantly anatomical and physiological (as in Foucault's analyses
of the microphysics of the prison or hospital), in control societies,
bodies are consonant with more distributed modes of individuation
that enable their infinite variation (informatic records, databases,
consumer profiles, genetic codes, identity shopping, workplace biometrics).[14] Their effects are network effects, and their agency is an
anonymous agency (in this sense, "anonymity" exists quite happily
alongside "identification").

This does not mean, however, that network control is simply
irrelevant, as if the mere existence of a network does away with the
notion of agency altogether. Network control ceaselessly teases out
elements of the unhuman within human-oriented networks. This is
most easily discovered in the phenomenology of aggregations in
everyday life: crowds on city streets or at concerts, distributed forms
of protest, and more esoteric instances of flashmobs, smartmobs, critical massing, or swarms of UAVs. All are different kinds of aggregations, but they are united in their ability to underscore the unhuman
aspects of human action. It is the unhuman swarm that emerges from
the genetic unit.

Network control is unbothered by individuated subjects (subjected
subjects). In fact, individuated subjects are the very producers and fa-
cilitators of networked control. Express yourself! Output some data!
It is how distributed control functions best.

*The twofold dynamic of network control—distributing agency while in-
stantiating rigid rules—implies that subjects acting in distributed networks
materialize and create protocols through their exercise of local agency.*

While Deleuze referred to it as "free-floating," control does not in
fact flit through the ether dissociated from real physical life. Quite
the opposite is true. Control is only seen when it materializes (though
in a paradoxical way), and it aims constantly to make itself "matter,"
to make itself relevant.

*In control societies, control "matters" through information—and infor-
mation is never immaterial.*[15]

Often control does this through bottom-up strategies that set the terms within which practices may possibly occur.[16] Network protocols are a bottom-up strategy, but at the end of the day, they exert massive control over technologies on a global scale.

Protocol in Computer Networks

It will be valuable at this point to explore further some of the aspects of actually existing networks with reference to two technoscientific systems, computer networks and biological networks. We hope this will underscore the material bent of the current approach.

Computer networks consist of nothing but schematic patterns describing various protocols and the organizations of data that constitute those protocols. These protocols are organized into layers. The Open System Interconnection (OSI) model, an abstract foundational model drafted in the 1980s for guiding the design of everything from secure private networks to normal Internet e-mail and Internet telephony, outlines seven layers for networked communication. Four of these seven are used in the design of most Internet communications: (1) the application layer (e.g., TELNET, the Web), (2) the transport layer (e.g., transmission control protocol [TCP]), (3) the Internet layer (e.g., Internet protocol [IP]), and (4) the physical layer (e.g., Ethernet).

Technical protocols are organized into layers (application, transport, Internet, physical); they formalize the way a network operates. This also allows us to understand networks such as the Internet as being more than merely technical.

These technical layers are nested, meaning that the application layer is nested within the transport layer, which is nested with the Internet layer, and so on. Each layer typically interfaces only with the layer immediately below or immediately above it. At each level, the protocol higher in precedence parses and encapsulates the protocol lower in precedence. Both actions are pattern based: on the one hand, parsing (computing checksums, measuring size, and so on) is about forcing data through various patterns, while on the other, encapsulation means adding a specific pattern of information to the beginning of the data object. For most Web traffic, the outermost layers are the IP layer and the TCP layer. Next typically comes an http header, which in turn encapsulates HTML text and simple ASCII text. Many technological protocols come into play during any typical network transaction, some interesting to humans, others interesting only to machines.

The application layer is perhaps most interesting to humans. It operates at the level of user software. The application layer often must deal with the messy requirements of human users, users who care about the semantic quality of "content."

A good metaphor for application layer communications is the perfunctory "paratextual" headers and footers attached to a written letter such as a salutation, a signature, the date, or a page number. These add-ons serve to encapsulate and structure the content of the letter, which itself is written using entirely different protocols (poetry, prose, or what have you). The application layer is unconcerned with infrastructural questions such as addressing or routing of messages. It simply frames and encapsulates the user "content" of the communication at the highest level.

The transport layer is the next layer in the hierarchy. The transport layer is responsible for making sure that data traveling across the network arrives at its destination correctly.

The transport layer acts as a concierge. It ensures that messages are bundled up correctly and are marked with the appropriate tags indicated by the various application layers encapsulated by it— e-mails directed over here, Web pages over there. In TCP, for example, each application in the application layer is inscribed into the transport header by numbers representing the source and destination ports (ports are computer interfaces that can send and receive data). To continue the metaphor, these are roughly equivalent to apartment numbers contained within a single building. If data is lost in transit, the transport layer is responsible for error correction. It is also the layer that is responsible for establishing persistent connections or "abstract circuits" between two machines.

The third layer is the Internet layer. This layer is more fundamental still than both the application and transport layers. The Internet layer is concerned with one thing: the actual movement of data from one place to another.

The Internet layer contains the source address of the machine sending the data, plus the destination address of the machine receiving the data. It is not aware of the type of data it is sending, simply the sender and receiver machines for that data. The Internet layer guides messages as they are routed through complex networks. It is also able to repackage the message in such a way as to fit through very small pipes or flow quickly through large ones.

The fourth layer, the physical layer, is the protocol layer specific to the actual material substrate of the communication network.

Copper wires have different physical properties from fiber-optic wires, despite the fact that both are able to transport an e-mail from one place to another. The physical-layer protocols interface directly with photons, electrons, and the material substrate, be it glass or metal or another medium, that allows them to flow. Consequently the physical layer is highly variable and differs greatly depending on the technology in question. It is less of a purely software-based layer in that it must take into account the material properties of the communications medium.

We wish to foreground the layer model of the Internet for several reasons. The first is to illustrate the technical basis for how multiple or "exceptional" topologies may coexist in the same network.

A classic example is the topological schism between the Domain Name System and the Internet protocol, two technologies that are intensely interconnected but are structured on radically different models of network control and organization: the Domain Name System is a database of information that is centralized in its core administration but decentralized in its global implementation (there are a limited number of top-level name servers, yet all subsequent name resolution is delegated down the chain to individual service providers and users), while the Internet protocol largely remains true to the radically distributed addressing and routing technologies proposed by scientists like Paul Baran, Donald Davies, and Leonard Kleinrock in the early 1960s. The distributed network topology, which Baran knew to be "exceptional" vis-à-vis the then-existing model of communication infrastructures dominated by AT&T's telephony network, is in some senses tempered by any number of more conventional (or sometimes reactionary) topologies that may exist in different layers. An example would be the deployment of digital rights management (DRM) usage restrictions within a piece of networked software. The DRM technologies exert centralized, coercive control from within the application layer even if they ultimately must burrow inside the TCP/IP layers to connect across the network. In this sense, network-based DRM shows how two antagonistic network topologies may work in coordination.

The "diachronic blindness" lamented earlier is also remedied somewhat with an investigation into how some of the core protocols deal with state changes and transformation over time.

For example, TCP is a state-based protocol, meaning that certain knowledge about the past is embedded in the technology itself. TCP-enabled network interfaces may be in one of several states. The network actions they perform will change their current state based on history and context. Being state based allows TCP to create a virtual circuit between sender and receiver and to perform actions such as error correction. The cruel irony is that history, context, and even "memory" have been exceedingly well integrated into any number of control technologies. Thus "trust" technologies that grew out of innovative social network research are now just as often deployed in data-mining and profiling operations related to national and international security. And likewise the "data trails" left by both human and nonhuman actors are the key data points for the compilation of profiles and the extraction of trends via the statistical analysis of their interconnection. Not all network technologies are state based, however, meaning that large sections of networking technologies act in a more synchronic manner.

The third point relates to the foregoing discussion about substance and individuation.

While networks are always material in the sense that they consist of material technologies such as electronic machines and physical media (metal, air, light, plastic), we are still reticent about using the philosophical model of substance to define a network. The process of instantiating and defining data is better understood as a process of individuation. All informatic sciences must deal with this issue. In computer science, certain artifices are used to "sculpt" undifferentiated data into discrete units or words, the most basic of which is the convention of collecting of eight binary bits into a byte. And beyond this, computer languages are designed with detailed technologies of individuation whereby specific mathematical values, such as a segment of memory, are given over to artificially designated types such as a character from A to Z or a decimal point number. Further up on the ladder of abstraction, generic data values may be "informed" or individuated into complex constructs such as data structures, objects, and files. The layer model of Internet communication is an extension of this privileging of modes of individuation over substance. The various layers are artificial and arbitrary from the perspective of raw bits of data; however, following all the allowances of the technologies involved, the "substance" of data is informed and individuated in specific, technologically intelligible ways. Indeed, many software exploits come from the voluntary transgression of individuated bounds. Thus a buffer overflow exploit "overflows" out of the expected boundaries of a given memory buffer writing to adjacent locations in memory and in doing so might cause the system to perform in a way it was not designed to.

Last is the principle of distributed sovereignty, the idea that control and organization are disseminated outward into a relatively large number of small, local decisions.

This process may be partial, as in the case of the Domain Name System, which, being decentralized, organizes a core subset of the technology into centralized systems while delegating the rest to local control. Or it may be more extensive, as in the case of IP routing, which uses a more anarchic, "emergent" model of decision making and control, whereby individual routers in the network make a number of local decisions that cumulatively result in robust networkwide functionality and "intelligence." Much of this design also flows from the so-called end-to-end principle governing much of network protocol design, which states that networks should remain neutral as to their uses and all machinic and user functionality not necessary for pure data transfer should be consigned to the "edges" of the network (i.e., personal computers and servers, rather than the various waypoints within the network). The agnostic quality of layer nesting—that a higher layer simply encapsulates a lower layer, manipulating it in certain mathematically agnostic ways such as computing a checksum or recording the size of its payload—is one of the core technological design principles that allows for the distributed model of sovereignty and control to exist.

Protocol in Biological Networks

In the example of computer networks, "protocol" is both a technical term and, as we've suggested, a way of describing the control particular to informatic networks generally.

While the example of Internet protocols may be viewed as a bona fide technology, protocols also inhere in the understanding of biological life. In turn, this informational, protocol-based understanding has led to the development of biotechnologies that take on a network form.

What is the "protocol" of biological networks? Since the mid-twentieth century, it has become increasingly common to speak of genes, proteins, and cells in terms of "information" and "codes." As historians of science point out, the informatic view of genetics has its roots in the interdisciplinary exchanges between cybernetics and biology during the postwar period.[17] Today, in the very concept of the Human Genome Project, of genetically tailored pharmaceuticals, of transgenics or GM foods, and of the ongoing preparedness against bioterrorism and emerging infectious disease, there is the notion of a genetic "code" that remains central to an understanding of "life" at the molecular level.

In biotechnology, all processes are, at some point or another, related back to what is often called the "flow of biological information" in the cell.

As one contemporary genetics textbook states, "a cell will use the biological information stored as a sequence of bases in DNA to produce the proteins necessary to the functioning of that cell."[18] Biotechnology—as the instrumental enframing of such processes— follows suit, harnessing the "natural" or biological processes of cells, proteins, and genes to manufacture drugs, therapies, "model organisms" for lab testing, and so on. While the scientific understanding of genetic processes has become increasingly sophisticated since the research on the structure of DNA in the 1950s, at the root of any university-level genetics education today is what Francis Crick called the "central dogma." The central dogma states that DNA in the cell nucleus makes a single-stranded RNA, and that RNA exits the cell nucleus, moving to the ribosome (a structure within the cell), where individual molecules are assembled from the RNA "template" into a beadlike string of amino acids, which then folds into a complex, three-dimensional structure known as a protein. Put simply, Crick's central dogma is that DNA makes RNA, which makes proteins, and proteins, of course, make us. In a 1958 paper, "On Protein Synthesis," Crick collated the research of the previous decade and formalized the role of DNA in the living cell as that of informational control: "Information means here the *precise* determination of sequence, either of bases in the nucleic acid or of the amino acid residues in the protein."[19] In a sort of revised Aristotelianism, Crick and his colleagues (Erwin Chargaff, Max Delbrück, George Gamow, Alexander Rich, James Watson, and others) approached the so-called coding problem from a number of angles. The informatic view of life was steadily gathering momentum.

Biotechnologies today have their origins in the interplay between cybernetics and biology. Molecular biologists working in the 1950s and 1960s located the control of biological molecules in a code and pattern, in an informational sequence of DNA, RNA, or amino acid molecules.

At the heart of this decidedly informatic view of biological and genetic "life" was a principle that came to be called "base pair complementarity." As we know, DNA is a double-stranded molecule in a shape of a helix, tightly coiled into larger structures inside the nucleus known as chromosomes. DNA, as we also know, is not only double stranded but arranged in a sequence of alternating molecules (one of four nitrogenous bases), one strand connected to the other via sugar and phosphate molecules, like rungs on a twisted ladder. Thus DNA has three molecular components: a sugar (deoxyribose), a phosphate group, and one of four nitrogenous bases (adenine, thymine, guanine, cytosine). Base pair complementarity states that, in the chemical bonding affinity of DNA's double helix, certain bases will always bond with each other: adenine with thymine (A-T), and cytosine with guanine (C-G). Thus, if one side of the double helix is known (ATACGT), the complementary side is also known (TATGCA). What set Crick and other scientists on the "coding problem" of the 1950s was this *sequential* property of DNA. That is, between DNA, RNA, and a fully assembled protein, there was assumed to be something common to them all, some principle of control that enables the recursive and highly specific production of needed molecules.

The concept of base pair complementarity was and still is central to biotechnology. Crick and his colleagues quickly understood that some combinatory process was responsible for generating a high degree of complexity (thousands of proteins) from apparent simplicity (sequences of four bases in DNA).

Biomolecules were quickly understood to be *informed matters*, containing either directly or indirectly the information needed for carrying out cellular processes. In molecular biology and genetics, such processes are commonly referred to as "biological control." These processes include gene expression (how a network of genes is switched on and off to produce proteins), cellular metabolism (how the components of enzymes and organelles transform "food" molecules into energy), and membrane signaling (the lock-and-key specificity of bringing molecules through a cell membrane). Together such processes are generally responsible for maintaining the living eukaryotic cell. And while it would be reductive to say that DNA "causes" each process, there is an implicit understanding in molecular biology of the central role that DNA—as genetic information—plays in each process.

Base pair complementarity not only implies an informatic approach to studying life but also implies a notion of biological control (gene expression, cellular metabolism, membrane signaling).

Indeed, this basic principle of base pair complementarity not only lies at the root of many "natural" processes in the cell but also drives many of the techniques and technologies that have become associated with biotechnology. In the early 1970s, when the first genetic engineering experiments were carried out on bacteria, it was base pair complementarity that enabled researchers to "cut" and "paste" segments of DNA in a precise manner, paving the way for recombinant DNA, genetically modified organisms (GMOs), transgenic animals, and the first biotech company (Genentech). In the early 1980s, when researchers at the Cetus Corporation developed a technology for rapidly copying large amounts of desired DNA segments (polymerase chain reaction, or PCR), base pair complementarity again served as the basic mechanism. And in the mid-1990s, when the first oligonucleotide microarrays (or "DNA chips") were marketed by companies such as Affymetrix, base pair complementarity formed the basis of the DNA-silicon hybrid chips. All these techniques are currently part and parcel of biotechnology as a set of practices, supplemented recently by the nascent field of "bioinformatics."

As an informatic principle, as a concept concerning "informed matters," base pair complementarity can operate across different material substrates, be it in the living cell, in a petri dish or test tube, or, more recently, in a computer.

The widespread use of computer databases (GenBank), Web-based gene-finding algorithms (BLAST), and automated genome sequencing computers demonstrates the principle of base pair complementarity *in silico*, in addition to the in vitro and in vivo.

In short, the increasing integration of cybernetics and biology has resulted in an informatic view of life that is also a view of life as a network ("biological control").

But it is when we see biotechnology in its instrumental, yet *nonmedical, nonbiological* context, that the "protocols" of biological control become the most evident. One such example is the nascent field of DNA computing, or biocomputing.[20]

The actual design and construction of a computer made of DNA takes these protocols or biological control into a whole new field of concern beyond the traditional distinctions of biology and technology.

While DNA computing is so new that it has yet to find its "killer app," it has been used in a range of contexts from cryptography to network routing or navigation problems to the handheld detection of biowarfare agents. The techniques of DNA computing were developed in the mid-1990s by Leonard Adleman as a proof-of-concept experiment in computer science.[21] The concept is that the combinatorial possibilities inherent in DNA (not one but two sets of binary pairings in parallel, A-T, C-G) could be used to solve specific types of calculations. One famous one is the so-called traveling salesman problem (also more formally called a "directed Hamiltonian path" problem): imagine a salesman who must go through five cities. The salesman can visit each city only once and cannot retrace his steps. What is the most efficient way to visit all five cities? In mathematical terms, the types of calculations are called "NP complete" problems, or "nonlinear polynomial" problems, because they involve a large search field that gets exponentially larger as the number of variables increases (five cities, each with five possible routes). For silicon-based computers, calculating all the possibilities of such problems can be computationally taxing. However, for a molecule such as DNA, the well-understood principle of base pair complementarity (that A always binds to T, and C always binds to G) makes for something like a parallel-processing computer, but a computer made out of enzymatic annealing of single strands of DNA rather than microelectrical circuits. One can "mark" a segment of any single-stranded DNA for each city (using gene markers or fluorescent dye), make enough copies to cover all the possibilities (using a PCR thermal cycler, a type of Xerox machine for DNA), and then mix them in a test tube. The DNA will mix and match all the cities into a large number of linear sequences, and quite possibly, one of those sequences will represent the most efficient solution to the "traveling salesman" problem.

As a mode of biological control, DNA computing generates a network, one constituted by *molecules* that are also *sequences*, that is, *matter* that is also *information*. The nodes of the network are DNA fragments (encoded as specific nodes A, B, C, D, etc.), and the edges are the processes of base pair binding between complementary DNA fragments (encoded as overlaps A-B, B-C, C-D, etc.). But DNA computing doesn't simply produce one network, for in solving such NP problems, it must, by necessity, produce many networks (most of which will not be a "solution"). The control mechanism of the DNA computer, therefore, relies on the identification and extraction of one subnetwork at the expense of all the others.

The network resulting from the experiment is actually a set of networks in the plural; the DNA computer generates a large number of networks, each network providing a possible path.

The network is therefore a series of DNA strands; it is combinatorial and recombinatorial. In addition, the networks produced in DNA computing move between the conceptual and the artifactual, between the ideality of mathematics and the materiality of the DNA computer's "hardware," moving from one medium (numbers, bits) to another, qualitatively different medium (DNA, GPCRs, citric acid cycle).[22]

DNA computing demonstrates protocological control at the micro-level of biomolecules, molecular bonds, and processes of annealing and denaturing. DNA computing shows how the problem-solving process does not depend on any one problem-solving "agent," but that the solution (mathematically and biochemically) arises from a context of distributed regulation. The solution comes not from brute number crunching but from an open, flexible array of total possibilities. This is how it is protocological.

The exponential search field provides the DNA computer with a context within which base pair complementarity operates in a highly distributed fashion. The protocological aspect of this system is not some master molecule directing all processes; it is immanent to the biochemical context of base pair annealing, functioning across the search field itself, which is in turn articulated via the technical design and implementation of the "network."

This means that DNA computing facilitates a peer-to-peer set of relationships between its nodes (base pairs) that bind or do not bind, given certain protocological parameters (complementarity, temperature, cycling chemical medium). From this perspective, DNA computing carries out its computations without direct, centralized control; all that the DNA computer requires is a context and a problem set defining a search field (such as the Hamiltonian path). However—and this too is protocological—this distributed character in no way implies a freedom from control. Recall that one of the primary concerns of the ARPANET was to develop a network that would be robust enough to survive the failure of one or more of its nodes. Adleman's Hamiltonian path problem could just as easily be reconceived as a contingency problem: given a directed path through a given set of nodes, what are the possible alternative routes if one of the nodes is subtracted from the set?

An Encoded Life

We have, then, two networks—a computer network and a biological network—both highly distributed, both robust, flexible, and dynamic.

While the first type of network (Internet protocols) is silicon based and may use biological concepts (intelligent agents, artificial life, genetic algorithms), the second (DNA algorithms) is fully biological and yet recodes itself in computational terms (biology as computation, as opposed to evolution).

Two "computers," two networks—two protocols? Yes and no. The example of DNA computing suggests that protocological control can be biological as well as computational. But what is the protocol? On the one hand, the aim of the experiment is mathematical and computational, yet on the other, the medium through which this is realized is biological and biochemical. So while computational protocols (flow control, data types, callback methods) may govern the inner workings of the informatic component of DNA computing, protocols also govern the interfacing between wet and dry, between the informatic and the biological. So there are two orders happening simultaneously. In the example of TCP/IP, protocological control is almost exclusively mathematical and computational, with the wetware being left outside the machine. Protocols facilitate the integration and standardization of these two types of networks: an "inter" network relating different material orders (silicon and carbon), and an "intra" network relating different variables within protocological functioning (nodes as DNA; edges as base pair binding). The protocol of biocomputing therefore does double the work. It is quite literally *bio*technical, integrating the logics and components specific to computers with the logics and components specific to molecular biology.

This is to emphasize a point made at the outset: protocol is a materialized functioning of distributed control. Protocol is not an exercise of power "from above," despite the blatant hierarchical organization of the Domain Name System or the rigid complementarity grammar of DNA. Protocol is also not an anarchic liberation of data "from below" despite the distributive organization of TCP/IP or the combinatory possibilities of gene expression. Rather, protocol is an immanent expression of control. Heterogeneous, distributed power relations are the absolute essence of the Internet network or the genome network, not their fetters. Thus the relation between protocol and power is somewhat inverted: the greater the distributed nature of the network, the greater the number of inside-out controls that enable the network to function as a network.

In both computer and biological networks, the primary function of protocol is to direct flows of information. In this sense, the networks we have described are not new but have their ontological foundations in cybernetics, information theory, and systems theory research in the mid-twentieth century.

This should come as no surprise, for both computer science and molecular biology have their common roots in World War II and postwar technical research.[23] The MIT mathematician and defense researcher Norbert Wiener defined information as a choice or selection from a set of variables. His influential book *Cybernetics, or Control and Communication in the Animal and the Machine* looked across the disciplines from electrical engineering to neurophysiology and suggested that human, animal, and mechanical systems were united in their ability to handle input/sensor and output/effector data in the ongoing management of the system. A central aspect to such cybernetic systems was the role of feedback, which implied a degree of self-reflexivity to any system.[24] Information, for Wiener, is a statistical choice from among the "noise" of the surrounding world, and as such it implies an apparatus with the ability to instantiate the very act of choice or selection.[25] Wiener refers to this ability as "control by informative feedback."

While Wiener was doing cybernetic military research on anti-aircraft ballistics, his colleague Claude Shannon was doing telecommunications research for Bell Labs. Much of Shannon's work with Warren Weaver is acknowledged as the foundation for modern telecommunications and can be said to have paved the way for the idea of the ARPANET in the late 1960s. Shannon's work, while much less interdisciplinary than Wiener's, resonated with cybernetics in its effort to define "information" as the key component of communications technologies (indeed, Wiener cites Shannon's work directly). Shannon and Weaver's information theory emphasized the quantitative view of information, even at the expense of all consideration of quality or content. As they state: "Information must not be confused with meaning. In fact, two messages, one of which is heavily loaded with meaning and the other of which is pure nonsense, can be exactly equivalent, from the present viewpoint, as regards information."[26] Such a hard-nosed technical view can still be seen today in the Internet's implementation of packet switching, in which chunks of data are fragmented and routed to destination addresses. While analysis of data packets on the Internet can be interpreted to reveal content, the technical functioning has as its implicit priority the delivery of quantity X from point A to point B.

If both cybernetics (Wiener) and information theory (Shannon) imply a quantitative, statistical view of information, a third approach, contemporaneous with cybernetics and information theory, offers a slight alternative.

The biologist Ludwig von Bertalanffy's "general systems theory" differs significantly from the theories of Wiener or Shannon. Wiener viewed human, animal, and mechanical systems together from an electrical engineering perspective, while Shannon viewed human users as separate from the communications technologies they used. By contrast, von Bertalanffy's work stressed the view of human or technological systems from a biological standpoint:

> The organism is not a static system closed to the outside and always containing the identical components; it is an open system in a quasi-steady state, maintained constant in its mass relations in a continuous change of component material and energies, in which material continually enters from, and leaves into, the outside environment.[27]

This view has several consequences. One is that while von Bertalanffy does have a definition of "information," it plays a much smaller role in the overall regulation of the system than other factors. Information is central to any system, but it is nothing without an overall logic for defining information and using it as a resource for systems management. In other words, the logics for the handling of information are just as important as the idea of information itself.

Another consequence is that von Bertalanffy's systems theory, in its organicist outlook, provides a means of understanding "information" in biological terms, something lacking in engineering or communications. This is not to suggest that systems theory is in any way more accurate or successful than the theories of Wiener or Shannon. What the genealogies of cybernetics, information theory, and systems theory show, however, is that "information," and an informatic worldview, display an ambivalent relation to the material world. On the one hand, information is seen as being abstract, quantitative, reducible to a calculus of management and regulation—this is the disembodied, immaterial notion of "information" referred to earlier. On the other hand, cybernetics, information theory, and systems theory all show how information is immanently material, configured into military technology, communications media, and even biological systems.

In the cybernetic feedback loop, in the communications channel of information theory, and in the organic whole of systems theory there exists a dual view of information as both immaterial and materializing, abstract and concrete, an act and a thing.

In short, we can say that what Deleuze calls "societies of control" provide a *medium* through which protocol is able to express itself. In such an instance, it is "information"—in all the contested meanings of the term—that constitutes the ability for protocol to materialize networks of all kinds. Protocol always implies some way of acting through information. In a sense, information is the concept that enables a wide range of networks—computational, biological, economic, political—to be networks. Information is the key commodity in the organizational logic of protocological control.

Information is the substance of protocol. Information makes protocol matter.

To summarize, if protocol operates through the medium of the network, and if information is the substance of that network, then it follows that the material effects of any network will depend on the establishment and deployment of protocol. While graph theory provides us with a set of useful principles for analyzing networks, we have seen that graph theory also obfuscates some core characteristics of networks: dynamic temporality, the lack of fixed node/edge divisions, and the existence of multiple topologies in a single network.

Toward a Political Ontology of Networks

We need an approach to understanding networks that takes into account their ontological, technological, and political dimensions. We will first restate the characteristics of protocol mentioned earlier: as a network phenomenon, protocol emerges through the complex relationships between autonomous, interconnected agents; protocological networks must be robust and flexible and must have material interfaces that can accommodate a high degree of contingency; protocological networks discriminate and regulate inclusively to their domain, not exclusively; principles of political liberalism guide all protocol development, resulting in an opt-in, total world system; and protocol is the emergent property of organization and control in networks that are radically horizontal and distributed. As we have shown, the "entity" in question may be the DNA computer and its laboratory techniques, or it may be the OSI Reference Model with its various layers for network protocols.

But if this is the case, we also need a set of concepts for interweaving the technical and the political. Ideally, our political ontology of networks would provide a set of concepts for describing, analyzing, and critiquing networked phenomena. It would depend on and even require a technical knowledge of a given network without being determined by it. It would view the fundamental relationships of control in a network as immanent and integral to the functioning of a network.

1. We can begin by returning to the concept of individuation, a concept that addresses the relation between the particular and the universal.

Individuation is a long-standing concept in philosophy, serving as the central debate among classical thinkers from Parmenides to Plato. Individuation is the process by which an entity is demarcated and identified as such. Individuation is different from the individual; it is a mobilization of forces that have as their ends the creation of individuals. In a sense, the question is how an individual comes about (which in turn leads to anxieties over causality). But individuation takes on new forms in the societies of control. As we've mentioned, for Deleuze a mode of individuation has little to do with individual human subjects, and more to do with the process through which a number of different kinds of aggregates are maintained over time. Individuation is key for understanding the construction of the entity, but it is equally key for understanding the construction and maintenance of the molecular aggregate (the network). Gilbert Simondon, writing about the relationships between individuation and social forms, suggests that we should "understand the individual from the perspective of the process of individuation rather than the process of individuation by means of the individual."[28]

A network deploys several types of individuation in the same time and space. It individuates itself as such from inside (organized political protests) or is individuated from the outside (repeated references by the United States to a "terrorist network").

The first type of individuation is that of the macroidentification of the network as a cohesive whole. This is, of course, a paradoxical move, since a key property of any network is its heterogeneity. Hence the first type of individuation is in tension with the second type of individuation in networks, the individuation of all the nodes and edges that constitute the system, for while the whole is greater than the sum of the parts, it is nevertheless the parts (or the localized action of the parts) that in turn constitute the possibility for the individuation of "a" network as a whole. Of course, the way the first individuation occurs may be quite different from the way the second occurs.

The individuation of the network as a whole is different from the individuation of the network components. However, both concern themselves with the topology of the network.

In the context of networks, individuation will have to be understood differently. Instead of the classical definition, in which individuation is always concerned with the production of individuals (be they people, political parties, or institutions), in the control society, individuation is always concerned with the tension between the individuation of networks as a whole and the individuation of the component parts of networks.

Individuation in the control society is less about the production of the one from the many, and more about the production of the many through the one. In the classical model, it is the hive that individuates the drone. Here, however, every drone always already facilitates the existence of multiple coexisting hives. It is a question not of being individuated as a "subject" but instead of being individuated as a node integrated into one or more networks. Thus one speaks not of a subject interpellated by this or that social force. One speaks instead of "friends of friends," of the financial and health networks created by the subject simply in its being alive.

The distinction from political philosophy between the individual and the group is transformed into a protocological regulation between the network as a unity and the network as a heterogeneity (what computer programmers call a "struct," a grouping of dissimilar data types). It is the management of this unity–heterogeneity flow that is most important. In terms of protocological control, the question of individuation is a question of how discrete nodes (agencies) and their edges (actions) are identified and managed as such. Identification technologies such as biometrics, tagging, and profiling are important in this regard, for they determine what counts as a node or an edge in a given network. Some key questions emerge: What resists processes of individuation? What supports or diversifies them? Does it change depending on the granularity of the analysis?

2. Networks are a multiplicity. They are robust and flexible.

While networks can be individuated and identified quite easily, networks are also always "more than one." Networks are multiplicities, not because they are constructed of numerous parts but because they are organized around the principle of perpetual inclusion. It is a question of a formal arrangement, not a finite count. This not only means that networks can and must grow (adding nodes or edges) but, more important, means that networks are reconfigurable in new ways and at all scales. Perhaps this is what it means to be a network, to be capable of radically heterogeneous transformation and reconfiguration.

In distributed networks (and partially in decentralized ones), the network topology is created by subtracting all centralizing, hermetic forces. The guerrilla force is a guerrilla force not because it has added additional foot soldiers but because it has subtracted its command centers.

"The multiple must be made," wrote Deleuze and Guattari, "not by always adding a higher dimension, but rather in the simplest of ways, by dint of sobriety, with the number of dimensions one already has available—always n − 1."[29] Like Marx's theory of primitive accumulation, it is always a question of inclusion through a process of removal or disidentification from former contexts. It is inclusion by way of the generic. The result is, as Deleuze argues, something beyond the well-worn dichotomy of the one and the many: "Multiplicity must not designate a combination of the many and the one, but rather an organization belonging to the many as such, which has no need whatsoever of unity in order to form a system."[30]

A technical synonym for multiplicity is therefore "contingency handling"; that is, multiplicity is how a network is able to manage sudden, unplanned, or localized changes within itself (this is built into the very idea of the Internet, for example, or the body's autoimmune system). A network is, in a sense, something that holds a tension within its own form—a grouping of differences that is unified (distribution versus agglomeration). It is less the nature of the parts in themselves, but more the conditions under which those parts may interact, that is most relevant. What are the terms, the conditions, on which "a" network may be constituted by multiple agencies? Protocols serve to provide that condition of possibility, and protocological control the means of facilitating that condition.

3. A third concept, that of movement, serves to highlight the inherently dynamic, process-based qualities of networks.

While we stated that networks are both individuated and multiple, this still serves only to portray a static snapshot view of a network.

Most of the networks we are aware of—economic, epidemiological, computational—are dynamic ones. Networks exist through "process," in Alfred North Whitehead's sense of the term, a "nexus" that involves a prehension of subject, datum, and form.

Perhaps if there is one truism to the study of networks, it is that networks are only networks when they are "live," when they are enacted, embodied, or rendered operational.

This applies as much to networks in their potentiality (sleeper cells, network downtime, idle mobile phones, zombie botnets) as it does to networks in their actuality. In an everyday sense, this is obvious— movements of exchange, distribution, accumulation, disaggregation, swarming, and clustering are the very stuff of a range of environments, from concentrated cities to transnational economies to cross-cultural contagions to mobile and wireless technologies. Yet the overwhelming need to locate, position, and literally pinpoint network nodes often obfuscates the dynamic quality of the edges. To paraphrase Henri Bergson, we often tend to understand the dynamic quality of networks in terms of stasis; we understand time (or duration) in terms of space. "There are changes, but there are underneath the change no things which change: change has no need of a support. There are movements, but there is no inert or invariable object which moves: movement does not imply a mobile."[31]

4. Finally, in an informatic age, networks are often qualified by their connectivity, though this is more than a purely technical term.

The peculiarly informatic view of networks today has brought with it a range of concerns different from other, non-IT-based networks such as those in transportation or analog communications. The popular discourse of cyberspace as a global frontier or as a digital commons, where access is a commodity, conveys the message that the political economy of networks is managed through connectivity. As Arquilla and Ronfeldt have commented, whereas an older model of political dissent was geared toward "bringing down the system," many current network-based political movements are more interested in "getting connected"—and staying connected.[32]

There are certainly many other ways of understanding networks akin to the ones mentioned here. We have tried to ground our views in an analysis of the actual material practice of networks as it exists across both the biological and information sciences.

We want to propose that an understanding of the control mechanisms within networks needs to be as polydimensional as networks are themselves.

One way of bridging the gap between the technical and the political views of networks is therefore to think of networks as continuously expressing their own modes of individuation, multiplicity, movements, and levels of connectivity—from the lowest to the highest levels of the network. In this way, we view networks as political ontologies inseparable from their being put into practice.

The Defacement of Enmity

Are you friend or foe? This is the classic formulation of enmity received from Carl Schmitt. Everything hinges on this relation; on it every decision pivots.

Friend-or-foe is first a political distinction, meaning that one must sort out who one's enemies are. But it is also a topological or *diagrammatic* distinction, meaning that one must also get a firm handle on the architectonic shape of conflict in order to know where one stands. Anticapitalism, for example, is not simply the hatred of a person but the hatred of an architectonic structure of organization and exchange. Friend-or-foe transpires not only in the ideal confrontation of gazes and recognitions or misrecognitions, as we will mention in a moment, but in the topological—that is, mapped, superficial, structural, and formal—pragmatics of the disposition of political force. To what extent are political diagrams and topologies of military conflict analogous to each other? On a simple level, this would imply a relationship between political and military enmity. For instance, first there is large-scale, symmetrical conflict: a standoff between nation-states, a massing of military force (front and rear regiments, waves of attack, the line of battle). Second, there is asymmetrical conflict: the revolution or insurgency that is a battle of maneuver targeted at what Clausewitz called the "decisive point" of vulnerability (flanking, surprise, multilinear attacks). From this, it is possible to identify a new type of symmetrical conflict today: decentralized and distributed operations across the political spectrum, from international terrorist networks to civil society protests to the latest military-technological operations (netwars or "network-centric warfare").

But here any congruency between the political and the topological starts to unravel, for as military historians note, a given conflict may display several politico-military topologies at once (for instance, the oscillation between massed forces and maneuver units in the armies of ancient Greece, the one personified in "angry" Achilles and the other in "wily" Odysseus), or a given topology of conflict may be adopted by two entirely incompatible political groups (the transnationals are networks, but so are the antiglobalization activists).

So a contradiction arises: the more one seeks to typecast any given political scenario into a given topological structure of conflict, the more one realizes that the raw constitution of political enmity is not in fact a result of topological organization at all; and the reverse is the same: the more one seeks to assign a shape of conflict to this or that political cause, as Deleuze and Guattari implicitly did with the rhizome, the more one sees such a shape as purely the agglomeration of anonymous force vectors, as oblivious to political expediency as the rock that falls with gravity or the bud that blooms in spring. But at the least, it is clear that one side of the congruency, the diagrammatic or topological, has gotten short shrift. What one may glean, then, from our notion of politico-military topologies is that there is a kind of materialism running through enmity, a kind of physics of enmity, an attempt to exercise a control of distributions (of people, of vehicles, of fire, of supplies).

Are you friend or foe? Everything depends on how one "faces" the situation; everything depends on where one is standing. Enmity is always a face because enmity is always "faced" or constituted by a confrontation. We stand alongside our friends; I stand opposite my foe. Friends only "face" each other insofar as they stand opposite and "face" their common foe (their enmity-in-common). Enmity is an interface.

The Schmittian friend-foe distinction is not just political in the military sense but political in the ethical sense, too. The basis of the friend-foe distinction is intimately related to the relation between self and other. But this self-other relation need not be a rapid fire of glances, gazes, and recognition (as in the Hegelian-Kojèvean model). For Levinas, ethics is first constituted by the ambiguous calling of the "face" of the other, for there is an affective dynamic at work between self and other that revolves and devolves around the "face" (a verb more than noun): "The proximity of the other is the face's meaning, and it means from the very start in a way that goes beyond those plastic forms which forever try to cover the face like a mask of their presence to perception."[33] Perhaps there is something to be learned by positioning Levinas in relation to Schmitt on this issue. A self does not set out, *avant la lettre*, to identify friend or foe according to preexisting criteria (political, military, ethical). Rather, the strange event of the "face" calls out to the self, in a kind of binding, intimate challenge: "The Other becomes my neighbor precisely through the way the face summons me, calls for me, begs for me, and in so doing recalls my responsibility, and calls me into question."[34] What implications does this have for enmity? Certainly that the self exists for the other. But it is much more than this. Enmity is not simply some final, absolute split (friend/foe, self/other) but rather a proliferation of faces and facings, whose very spatiality and momentary, illusory constrictions create the conditions of possibility for enmity in both its political and military forms (facing-allies, facing-enemies). Friends, foes, selves—there are faces everywhere.

But what of an enmity without a face? What of a defacement of enmity?

This is where a consideration of politico-military topologies comes into focus. Enmity is dramatized or played out in the pragmatic and material fields of strategy and war. The emerging "new symmetry" mentioned in our prolegomenon appears in a variety of forms: information-based military conflict ("cyberwar") and nonmilitary activity ("hacktivism"), criminal and terrorist networks (one "face" of the so-called netwar), civil society protest and demonstration movements (the other "face" of netwar), and the military formations made possible by new information technologies (C4I operations [command, control, communications, computers, and intelligence]). What unites these developments, other than that they all employ new technologies at various levels?

For Arquilla and Ronfeldt, it is precisely the shapeless, amorphous, and faceless quality that makes these developments noteworthy, for the topologies of netwar and the "multitude" throw up a challenge to traditional notions of enmity: they have no face; they are instances of faceless enmity. Or rather, they have defaced enmity, rendered it faceless, but also tarnished or disgraced it, as in the gentleman's lament that asymmetrical guerrilla tactics deface the honor of war.

These examples are all instances of *swarming,* defined as "the systematic pulsing of force and/or fire by dispersed, internetted units, so as to strike the adversary from all directions simultaneously."[35] Though it takes inspiration from the biological domain (where the study of "social insects" predominates), Arquilla and Ronfeldt's study of swarming is a specifically politico-military one. A swarm attacks from all directions, and intermittently but consistently—it has no "front," no battle line, no central point of vulnerability. It is dispersed, distributed, and yet in constant communication. In short, it is a faceless foe, or a foe stripped of "faciality" as such. So a new problematic emerges. If the Schmittian notion of enmity (friend-foe) presupposes a more fundamental relation of what Levinas refers to as "facing" the other, and if this is, for Levinas, a key element to thinking the ethical relation, what sort of ethics is possible when the other has no "face" and yet is construed as other (as friend or foe)? What is the shape of the ethical encounter when one "faces" the swarm?

A key provocation in the "swarm doctrine" is the necessary tension that appears in the combination of formlessness and deliberate strategy, emergence and control, or amorphousness and coordination.

As a concept, swarming derives from biological studies of social insects and their capacity to collectively carry out complex tasks: the construction of a nest by wasps, the coordinated flashing among fireflies, the cooperative transportation of heavy loads by ants, the global thermoregulation of a nest by bees, and so on.[36] Each of these examples illustrates the basic rules of biological self-organization, how a set of simple, local interactions culminates in complex, collective organization, problem solving, and task fulfillment. In philosophy, the relationship between humans and insects far predates such scientific studies: Aristotle, in calling the human being a "political animal," made a direct contrast to the insect; Hobbes stressed the human capacity to emerge out of the "state of nature" through the giving up of rights to a sovereign, a capacity lacking in insects; Marx acknowledged the capacity of insects to build but emphasized the human capacity for idealization as part of their "living labor"; and Bergson, expressing a poetic wonder at the ability of insects to self-organize, suggested that evolution was multilinear, with human beings excelling at "reason" and insects excelling at "intuition."

There is an entire genealogy to be written from the point of view of the challenge posed by insect coordination, by "swarm intelligence." Again and again, poetic, philosophical, and biological studies ask the same question: how does this "intelligent," global organization emerge from a myriad of local, "dumb" interactions?

Arquilla and Ronfeldt also ask this question, but they limit their inquiry to military applications. Building on the military-historical work of Sean Edwards, their analysis is ontological: it is about the relationship between *enmity* and *topology*. They distinguish between four types of aggregate military diagrams: the chaotic melee (in which person-to-person combat dominates, with little command and control), brute-force massing (where hierarchy, command formations, and a battle line predominate), complex maneuvers (where smaller, multilinear, selective flanking movements accompany massing), and finally swarming (an "amorphous but coordinated way to strike from all directions").[37] While elements of each can be found throughout history (horse-mounted Mongol warriors provide an early swarming

example), the interesting thing about swarming is the nagging tension in being "amorphous but coordinated." How is it possible to control something that is by definition constituted by its own dispersal,
by being radically distributed, spread out, and horizontal?

*Answering this question in the context of conflict (military or civilian)
means addressing the question of enmity. That is, if "control" in conflict is
ordinarily situated around a relationship of enmity (friend-foe, ally-enemy),
and if this relation of enmity structures the organization of conflict (symmetrical standoff, insurgency, civil disobedience), what happens when enmity
dissolves in the intangible swarm?*

In part this is the question of how conflict is structured in terms of
more complex modes of enmity ("going underground," "low-intensity
conflict," the "war on terror"). Are the terms of enmity accurate for
such conflicts? Perhaps it is not possible for a network to be an *enemy*?
Without ignoring their political differences, is there a topological
shift common to them all that involves a dissolving or a "defacing" of
enmity? Can a swarm be handled? If there is no foe to face, how does
one face a foe? It is not so much that the foe has a face, but that the
foe is faced, that "facing" is a process, a verb, an action in the making.
This is Levinas's approach to the ethical encounter, an encounter
that is based not on enmity but on a "calling into question" of the
self. But in a different vein, it is also the approach of Deleuze and
Guattari when they speak of "faciality." Not unlike Levinas, they stress
the phenomenal, affective quality of "facing." But they also take "facing" (facing the other, facing a foe) to be a matter of pattern recognition, a certain ordering of holes, lines, curves: "The head is included
in the body, but the face is not. The face is a surface: facial traits,
lines, wrinkles; long face, square face, triangular face; the face is a
map, even when it is applied to and wraps a volume, even when it
surrounds and borders cavities that are now no more than holes."[38]
Faciality is, in a more mundane sense, one's recognition of other human faces, and thus one's habit of facing, encountering, meeting others all the time. But for Deleuze and Guattari, the fundamental process
of faciality also leads to a deterritorialization of the familiar face, and
to the proliferation of faces, in the snow, on the wall, in the clouds,
and in other places (where faces shouldn't be) . . .

Places where faces shouldn't be—can this be what swarming is?

Or must one extract a "faciality" in every site of enmity? Consider an example from popular culture. In *The Matrix Revolutions*, there are two types of swarms, the first being the insectlike sentinels who attack the human city of Zion from all directions. In a textbook case of military swarming, they eventually defeat the humans' defensive by amassing scores of individual sentinels into one large, anthropomorphic face—a literal facialization of enmity. What started as a swarm without a face becomes a face built out of the substrate of the swarm. The Matrix appears to be at once totally distributed and yet capable of a high degree of centralization (swarm versus face). While earlier science fiction films could only hint at the threatening phenomenon of swarming through individual creatures (e.g., *Them!*), the contemporary science fiction film, blessed with an abundant graphics technology able to animate complex swarming behaviors down to the last detail, still must put a "face" on the foe, for in the very instant the swarm reaches the pinnacle of its power, its status as a defaced enemy is reversed and the swarm is undone. (*Tron* does something similar: the denouement of facialization comes precisely at the cost of all the various networked avatars zipping through the beginning and middle of the film.) Again the point is not that faciality—or cohesion, or integrity, or singularity, or what have you—is the sole prerequisite for affective control or organization, for indeed the swarm has significant power even before it facializes, but that faciality is a particular instance of organization, one that the swarm may or may not coalesce around. The core ambiguity in such expressions of swarming is precisely the tension, on the macro scale, between amorphousness and coordination, or emergence and control. Does coordination come on the scene to constrain amorphousness, or does it instead derive from it? Is a minimal degree of centralized control needed to harness emergence, or is it produced from it?

While the biological study of self-organization seems caught on this point, the politico-military-ethical context raises issues that are at once more concrete, more troubling, and more "abstract." In a sense, the swarm, *swarming-as-faciality*, is a reminder of the *defacement* proper not only to distributed insects but also to distributed humans;

swarming is simply a reminder of the defacement that runs through all instances of "facing" the other.

"The face is produced only when the head ceases to be part of the body."[39]

Biopolitics and Protocol

The discussion of swarming and networking hints again at a point made throughout this book: that the preponderance of network forms of organization in no way implies an absence of power relations or of control structures. In fact, it prescribes them. But the kinds of power relations and control structures are not necessarily those of top-down hierarchies. They are of another order.

Additionally, we should stress that all this talk about networks is not restricted to the discussion of technical systems. Networks can be technological, yes, but they are also biological, social, and political. Thus what is at stake in any discussion of the political dimensions of networks is, at bottom, the experience of living within networks, forms of control, and the multiple protocols that inform them. As we suggest, networks are always "living networks," and often what is produced in living networks is social life itself. That is, if there is something that results from networks, something produced from networks, it is the experience of systematicity itself, of an integration of technology, biology, and politics. Networks structure our experience of the world in a number of ways, and what is required is a way of understanding how networks are related to the aggregates of singularities—both human and nonhuman—that are implicated in the network. What is required, then, is a way of understanding politics as biopolitics.

Michel Foucault calls "biopolitics" that mode of organizing, managing, and above all regulating "the population," considered as a biological, species entity.

Broadly speaking, biopolitics is a historical condition in which biology is brought into the domain of politics; indeed, it is the moment at which "life itself" plays a particular role in the ongoing management of the social and the political.

This is differentiated from, but not opposed to, an "anatomo-politics," in which a range of disciplinary techniques, habits, and movements—most often situated within institutions such as the prison, the hospital, the military, and the school—collectively act on and render docile the individualized body of the subject.[40]

Contemporary interpretations of biopolitical instances vary from the medical-sociological emphasis on normativity, to the philosophical emphasis on the concept of "bare life," to the political and existential emphasis on biopolitics as the production of the social.[41]

Our approach will be very specific. We want to pose the question: what does biopolitics mean today, in the context of networks, control, and protocol?

If networks permeate the social fabric, and if networks bring with them novel forms of control, then it follows that an analysis of networks would have to consider them as living networks (and politics as biopolitics). Therefore we can consider the biopolitics of networks by highlighting two aspects nascent within Foucault's concept: that of biology and that of informatics.

Biopolitics is, as the term implies, an intersection between notions of biological life and the power relations into which those notions are stitched.

Both are historical, and both are constantly undergoing changes. Foucault mentions the regulation of birth and death rates, reproduction, pathology, theories of degeneracy, health and hygiene, as well as new statistical methods of tracking the migration of peoples within and between nations, all of which emerge from the end of the eighteenth century through the late nineteenth.[42] Biopolitical forms thus exist alongside the emergence of Darwinian evolution, germ theory, social Darwinism, and early theorizations of eugenics.

But biopolitics is not simply biology in the service of the state. It is created, in part, by a set of new technologies through which populations may be organized and governed. The accumulation and ordering of different types of information are thus central to biopolitics.

The development of the fields of statistics and demography was crucial for the development of modern biopolitics, what Foucault refers to as "sciences of the state." Thus the population is articulated in such a way that its totalization—and thus control—may be achieved via methods or techniques of statistical quantification. This is not only a mode of efficiency; it also forms a political economy of mass subjects and mass bodies that was to inform political economists from David Ricardo to Adam Smith.

Biopolitics, in its Foucauldian formulation, can thus be defined as the strategic integration of biology and informatics toward the development of techniques of organization and control over masses of individuals, species groups, and populations.

Foucault's biopolitics functions via three directives, each associated with today's network diagrams.

1. To begin with, biopolitics defines a specific object of governance: "the population."

Foucault notes a gradual historical shift from the disciplining of individualized bodies to the governance of populations. "Population" does not just mean the masses, or groups of people geographically bound (that is, "population" is not the same as "individual" or "territory"). Rather, the population is a flexible articulation of individualizing and collectivizing tendencies: many individual nodes, clustered together. Above all, the population is a political object whose core is biological: the population is not the individual body or organism of the subject-citizen but rather the mass body of the biological species.[43] In Foucault's context of the emergence of the modern state-form, the existence of the state is consonant with the "health" of the state. The main issue of concern is therefore how effectively to control the coexistence of individuals, groups, and relations among them.[44]

2. Biopolitics defines a means for the production of data surrounding its object (the population).

A crucial challenge is to determine how data pertinent to the regulation of the health of the population may be extracted from it. Within biomedicine, the "vital statistics" of reproduction, mortality rates, the spread of diseases, heredity, and other factors become a means for monitoring the population as a single, dynamic entity. This is evident in the recent events surrounding anthrax bioterrorism and the SARS epidemic, but it already exists in the push toward the computerization of health care ("infomedicine") and the use of information technologies to mediate doctor-patient relationships ("telemedicine"). As Foucault notes, in the nineteenth century, the use of informatic sciences in biopolitics enabled the implementation of new techniques for regulating the population. The increased use of statistics created changes in birth-control practices, as well as epidemiology, public hygiene campaigns, and criminology.

Taken together, the two elements of biology and informatics serve to make biopolitical control more nuanced, and more effective.

Confinement is on the wane, but what takes its place is control. "Our present reality," wrote Deleuze for a colloquium on Foucault, "presents itself in the apparatuses of *open and continuous control*."[45] It is not a contradiction to say that in societies of control there is both an increase in openness and an increase in control. The two go hand in hand. A new aggregation, the "new multiple body" (we can say a network), exists today, and regulation of this species-population would have to take into account how local actions affect global patterns of control. As Foucault notes, the characteristics of the population "are phenomena that are aleatory and unpredictable when taken in themselves or individually, but which, at the collective level, display constants that are easy, or at least possible, to establish."[46] Thus if any single node experiences greater freedom from control, it is most likely due to a greater imposition of control on the macro level. At the macro level, the species-population can not only be studied and analyzed but can also be extrapolated, its characteristic behaviors projected into plausible futures (birth/death rates, growth, development, etc.). The proto–information sciences of demographics, and later statistics, provided a technical ground for a more refined, mathematically based regulation and monitoring of the population.

However, far from a reductivist, homogenizing strategy, this modern regulative form of governmentality is predicated, for Foucault, on a dual approach, which both universalizes as well as individualizes the population. Populations can exist in a variety of contexts (defined by territory, economic and class groupings, ethnic groupings, gender-based divisions, or social institutions)—all within a framework analyzing the fluxes of biological activity characteristic of the population. Such groupings do not make sense, however, without a means of defining the criteria of grouping, not just the individual subject (and here "subject" may also be a verb) but a subject that could be defined in a variety of ways, marking out the definitional boundaries of each grouping. As individuated subjects, some may form the homogeneous core of a group, others may form its boundaries, its limit cases.

The methodology of biopolitics is therefore informatics, but a use of informatics in a way that reconfigures biology as an information resource. In contemporary biopolitics, the body is a database, and informatics is the search engine.

One important result of this intersection of biology and informatics in biopolitics is that the sovereign form of power (the right over death and to let live) gives way to a newer "regulative power" (the right to make live and let die). In other words, biology and informatics combine in biopolitics to make it productive, to impel, enhance, and optimize the species-population as it exists within the contexts of work, leisure, consumerism, health care, entertainment, and a host of other social activities.

3. After defining its object (the biological species-population) and its method (informatics/statistics), biopolitics reformulates the role of governance as that of real-time security.

If the traditional sovereign was defined in part by the right to condemn to death, biopolitics is defined by the right to foster life. The concern is not safety anymore, but security. Safety is about being removed from danger, being apart and surrounded by walls and protections of various kinds, and hence is a fundamentally modern notion; but security means being held in place, being integrated and immobile, being supported by redundant networks of checks and backups, and hence is a thoroughly information-age idea. Security is as much an economic and cultural notion as it is a military one. (Certainly the most evident militaristic example is the recent scare over terrorism and the budgetary boost in the U.S. biowarfare research program.) Biopolitics also remains consonant with neoliberalism in its notion of humanitarian security in the form of health insurance, home care, outpatient services, and the development of biological "banking" institutions (sperm and ova banks, blood banks, tissue banks, etc.). The computer scientists are hawkish on security, too. "Above all," the Internet standards community writes, "new protocols and practices must not worsen overall Internet security. A significant threat to the Internet comes from those individuals who are motivated and capable of exploiting circumstances, events, or vulnerabilities of the system to cause harm."[47] In fact, security is so central to networked organization that, as a rule, each new Internet protocol must contain a special written section devoted to the topic.

In biopolitics, security is precisely this challenge of managing a network of technologies, biologies, and relations between them. Security can be defined, simply, as the most efficient management of life (not necessarily the absence of danger or some concept of personal safety).

The challenge of security is the challenge of successfully fostering a network while maintaining an efficient hold on the boundaries that the networks regularly transgress. This is seen in a literal sense in cases surrounding computer viruses (firewalls) and epidemics (travel advisories). The problem of security for biopolitics is the problem of creating boundaries that are selectively permeable. While certain transactions and transgressions are fostered (trade, commerce, tourism), others are blockaded or diverted (sharing information, the commons, immigration). All these network activities, many of which have

become routine in technologically advanced sectors of the world, stem
from a foundation that is at once biological and technological.

Despite these aspects of biopolitics, the relation between life and
politics within a technoscientific frame is never obvious, or stable. In
both the sovereign and liberal democratic formations, we find what
Giorgio Agamben calls a "zone of indistinction" between "bare life"
(biological life) and the qualified life of the citizen (the political subject).[48] In more contemporary situations, we find what Michael Hardt
and Antonio Negri refer to as "living immaterial labor" (the "intellec-
tual labor" of the IT industries, technoscience, and media industries).[49]
Biopolitics is therefore never an ideology but rather is a particular
problem emerging within protocological relations about how and
whether to distinguish "life" from "politics."

At this point, we can summarize some of the main features of the
biopolitical view of networks in roughly the following way:

- A shift in the object of control, from the individualized body
  or organism of the civil subject to the massified biological
  species-population.
- The development and application of informatic technologies
  such as statistics and demographics, for calculating averages,
  generating hierarchies, and establishing norms.
- A change in the nature of power, from the sovereign model
  of a command over death to a regulatory model of fostering,
  impelling, and optimizing life.
- The effect of making the concern for security immanent to
  social life while also creating networks of all types (social,
  economic, biological, technological, political).[50]

What is the relationship between the Foucauldian concept of biopolitics and our constellation of topics: networks, control, and protocol?
We can start by saying that in the way in which it operates, biopolitics conceives of networks as existing in real time and therefore as
existing dynamically. Biopolitics is a mode in which protocological
control manages networks as living networks. In the biopolitical perspective, we perceive a strategic indiscernibility between an informatic
view of life and the notion of "life itself." The concept of "life itself"

is central in the development of biological science, but its meanings are not limited to the biological. It is both essentialist (the most real, immediate foundation) and constructed (new sciences such as genetics provide new views of "life itself"). "Life itself" is constructed as a type of limit point, the degree zero of organic vitality. The concept is constructed as having nothing behind or beneath it. "Life itself" is thought to be self-evident as a concept, and therefore it often remains unquestioned. Yet we suggest that, like any ideological construct, the concept of "life itself" must continually be produced and reproduced in order to retain its power as a meme. This is evidenced not only by examples in biomedicine (DNA as the unimpeachable "book of life") but also in the social existence of online virtual communities, wireless mobile technologies, and e-consumerism and entertainment, which take as their target the elevation of an aestheticized life or "lifestyle."

Biopolitics can be understood, in our current network society, as the regulation—and creation—of networks as living networks.

Biopolitics achieves this through a multistep process: first, all living forms must be made amenable to an informatic point of view (the hegemony of molecular genetics and cybernetics plays an important role here). Then, once "information" can be viewed as an index into "life," that information is accommodated by network structures (algorithms, databases, profiles, registrations, therapies, exchanges, login/password). Finally, once life is information, and once information is a network, then the network is made amenable to protocols—but with the important addition that this real-time, dynamic management of the network is also a real-time, dynamic management of "life itself" or living networks.[51]

Life-Resistance

The stress given to security in recent years has made it more and more difficult to offer a compelling theory of resistance, one unfettered by the dark epithets so easily crafted by the political Right—such as "cyberterrorist" or even "hacker," a word born from love and now tarnished by fear—but we would like to try nonetheless.

*The target of resistance is clear enough. It is the vast apparatus of techno-
political organization that we call protocol.*

Deleuze described it as "new forms of resistance against control societies."[52] Protocological control is key because protocol implements
the interactions between networked nodes in all their minute detail.
But, as we discussed previously, these protocological networks foster
the creation and regulation of life itself. In other words, the set of
procedures for monitoring, regulating, and modulating networks as
living networks is geared, at the most fundamental level, toward the
production of life, in its biological, social, and political capacities. So
the target is not simply protocol; to be more precise, the target of
resistance is the way in which protocol inflects and sculpts life itself.

Power always implies resistance. In relationships of force there are
always vector movements going toward and against. This is what
Hardt and Negri call "being-against."[53] It is the vast potential of human life to counter forces of exploitation. Contemplating this in the
context of intranetwork conflict, we can ask a further question: How
do networks transform the concept of resistance? Can the exploited
become the exploiting? As we've stated, the distributed character of
networks in no way implies the absence of control or the absence of
political dynamics. The protocological nature of networks is as much
about the maintenance of the status quo as it is about the disturbance
of the network. We can begin to address this issue by reconsidering
resistance within the context of networked technology.

*If networks are not just technical systems but also real-time, dynamic,
experiential living networks, then it makes sense to consider resistance as
also living, as "life-resistance."*

There are (at least) two meanings of the phrase "life-resistance":
(1) life is what resists power, and (2) to the extent that it is co-opted
by power, "life itself" must be resisted by living systems. For the first
definition (life as resistance to power), we are indebted to Deleuze,
who in turn drew on Foucault. "When power . . . takes life as its aim
or object," wrote Deleuze, "then resistance to power already puts

itself on the side of life, and turns life against power."[54] Confirming the direct link between power and life, he continues: "Life becomes resistance to power when power takes life as its object."[55] And then, for maximum clarity, Deleuze states this point a third time: "When power becomes bio-power, resistance becomes power of life, a vital-power that cannot be confined within species, places, or the paths of this or that diagram.... Is not life this capacity to resist force?... there is no telling what man might achieve 'as a living being,' as the set of 'forces that resist.'"[56]

In this sense, life is the capacity to resist force.

It is the capacity to "not become enamored of power," as Foucault put it years earlier in his preface to the English edition of Deleuze and Guattari's *Anti-Oedipus*. Life is a sort of counterpower, a return flow of forces aimed backward toward the source of exploitation, selectively resisting forms of homogenization, canalization, and subjectification. (But then this is really not a resistance at all but instead an intensification, a lubrication of life.)

On the other hand, life is also that which is resisted (resistance-to-life), that against which resistance is propelled.

Today "life itself" is boxed in by competing biological and computational definitions. In the biological definition, the icon of DNA is mobilized to explain everything from Alzheimer's to attention deficit disorder. In the computational definition, information surveillance and the extensive databasing of the social promote a notion of social activity that can be tracked through records of transactions, registrations, and communications.[57] Resistance-to-life is thus a challenge posed to any situation in which a normative definition of "life itself" dovetails with an instrumental use of that definition. Aiming toward a transformative, bioethical goal, the resistance-to-life perspective poses the question: Whom does this definition of life benefit and how? It asks: Is it enough to define life in this way? What other types of life are shut down by this definition of life? Here is the critic D. N. Rodo-wick writing on the theme of resistance-to-life in Deleuze:

Resistance should be understood, then, as an awakening of forces of life that are more active, more affirmative, richer in possibilities than the life we now have.

For Foucault, the stakes of a bio-power mean that life must be liberated in us because it is in us that life is imprisoned.... [The individual is marked], once and for all, with a known and recognizable identity—you will be White or Black, masculine or feminine, straight or gay, colonizer or colonized, and so on. Alternately, resistance means the struggle for new modes of existence. It is therefore a battle for difference, variation, and metamorphosis, for the creation of new modes of existence.[58]

Thinking beyond identity politics, we can easily expand Rodowick's list to include the instrumental definitions of life that abound today in the pharmaceutical industry, in biological patents, in monocultural crops, and in genetic reductionism.

The point for resistance-to-life, then, is not simply to resist "you will be X or Y," but also to resist the ultimatum "you will be sick or healthy" and all the other yardsticks of the pharmaco-genetic age.

Resistance-to-life is that which undoes the questionable formulation of "life itself" against its underpinnings in the teleology of natural selection and the instrumentality of the hi-tech industries, and against the forms of social and political normalization that characterize globalization. Resistance-to-life is specifically the capacity to resist the fetishization of life and its reification into "life itself."

Life-resistance is not in any way an essence, for this would amount to saying that life itself is spiritual or transcendental. Life-resistance is not a vitalism, either (at least not in its nineteenth-century formulation of a metacorporeal life force), but neither is it a social constructionism (and the complete relativism of "life itself" that follows from this).

Life-resistance is nothing more than the act of living. It is an inductive practice and proceeds forward through doing, thereby going beyond "life itself." Life-resistance is, above all, a productive capacity. In this sense, life-resistance is rooted in what Henri Bergson called the élan vital or what Deleuze and Guattari called the machinic phylum.

Perhaps what is most instructive about this view of biopolitics and resistance is that life-resistance is not exclusive to *human* agencies and actions, especially when considered from the perspective of networks as living networks. Life-resistance puts forth the difficult, sometimes frustrating proposition that "life" is not always synonymous with the limited cause-and-effect relations usually attributed to human agencies; in this sense, networks—or living networks—contain an anonymity, a nonhuman component, which consistently questions common notions of action, causality, and control.

The Exploit

In the non-protocological arenas, progressive political change is generated through struggle, through the active transfer of power from one party to another. For example, the institution of the forty-hour workweek was the result of a specific shift in power from capital to labor. To take another example, women's liberation is the result of specific transfers of power in the areas of law (suffrage, abortion, birth control), in the expectations surrounding domestic labor, biological and social ideas about gender, and so on.

Yet within protocological networks, political acts generally happen not by shifting power from one place to another but by exploiting power differentials already existing in the system.

This is due mainly to the fundamentally informatic nature of networks. Informatic networks are largely immaterial. But immaterial does not mean vacillating or inconsistent. They operate through the brutal limitations of abstract logic (if/then, true or false).

Protocological struggles do not center around changing existent technologies but instead involve discovering holes in existent technologies and projecting potential change through those holes.[59] *Hackers call these holes "exploits."*

Thinking in these terms is the difference between thinking socially and thinking *informatically*, or the difference between thinking in terms of probability and thinking in terms of *possibility*. Informatic spaces do not bow to political pressure or influence, as social spaces do. But informatic spaces do have bugs and holes, a by-product of high levels of technical complexity, which make them as vulnerable to penetration and change as would a social actor at the hands of more traditional political agitation.

Let us reiterate that we are referring only to *protocological* resistance and in no way whatsoever suggest that non-protocological practice should abandon successful techniques for effecting change such as organizing, striking, speaking out, or demonstrating. What we suggest here is a supplement to existing practice, not a replacement for it.

The goal for political resistance in life networks, then, should be the discovery of exploits—or rather, the reverse heuristic is better: look for traces of exploits, and you will find political practices.

Let's flesh out this idea using examples from actual practice, from specific scenarios. The first is an instance of the protocological masquerading as biological: the computer virus. Deleuze mentions computer viruses in his 1990 interview with Negri:

> It's true that, even before control societies are fully in place, forms of delinquency or resistance (two different things) are also appearing. Computer piracy and viruses, for example, will replace strikes and what the nineteenth century called "sabotage" ("clogging" the machinery).[60]

Computer viruses have a spotted history; they often involve innovative programming techniques that have been used in other areas of computer science, but they are also often tagged as being part of delinquent or criminal activities. Should computer viruses be included in the "history" of computers? How much have viruses and antivirus programs contributed to the development of "official" computer science and programming? The majority of the early instances of computer viruses have ties to either the university or the corporation: the "Darwin" game (AT&T/Bell Labs, early 1960s), "Cookie Monster" (MIT, mid-1960s), "Creeper" and "Reaper" (BBN, early 1970s), "tapeworm" (XeroxPARC, early 1970s), and so on.[61] Like early hacking activities, their intent was mostly exploratory. Unlike hacking, however, the language of biology quickly became a provocative tool for describing these encapsulations of code. Science fiction classics such as John Brunner's *The Shockwave Rider* popularized the vitalism of computer viruses, and by the early 1980s, researchers such as Fred Cohen published articles on "computer viruses" in academic journals such as *Computers and Security*.[62]

In terms of understanding networks, one of the greatest lessons of computer viruses and their cousins (Internet worms, Trojan horses) is that, like biological viruses, they exploit the normal functioning of their host systems to produce more copies of themselves. Viruses are life exploiting life.

Standard computer viruses essentially do three things: they "infect" documents or a program by overwriting, replacing, or editing code; they use the host system to create copies and to distribute copies of themselves; and they may have one or more tactics for evading detection by users or antivirus programs.

Contrary to popular opinion, not all computer viruses are destructive (the same can be said in biology, as well). Certainly computer viruses can delete data, but they can also be performative (e.g., demonstrating a security violation), exploratory (e.g., gaining access), or based on disturbance rather than destruction (e.g., rerouting network traffic, clogging network bandwidth). Originally computer viruses operated in one computer system at a time and required external media such as a floppy disk to pass from computer to computer. The increasing popularity of the civilian Internet, and later the Web, made possible network-based, self-replicating computer viruses (or "worms"). For instance, in January 2003, the "Sobig.F" virus, which used e-mail address books, generated an estimated one hundred million "fake" e-mails (which amounted to one in seventeen e-mails being infected in a given day). That same year, the "Blaster" virus rapidly spread through the Internet, infecting some four hundred thousand operating systems—all running Microsoft Windows. An attempt was made to release an automated anti-Blaster "vaccine" called "Naachi"; but the attempt failed, as the purported fix clogged networks with downloads of the Blaster patch from the Microsoft Web site (temporarily disabling a portion of the Air Canada and the

U.S. Navy computer clusters). Over the years, a whole bestiary of infectious code has come to populate the information landscape, including spam, spyware, and adware, as well as other nonvirus automated code such as intelligent agents, bots, and webcrawlers.

Computer viruses thrive in environments that have low levels of diversity.

Wherever a technology has a monopoly, you will find viruses. They take advantage of technical standardization to propagate through the network. (This is why Microsoft products are disproportionately infected by viruses: Microsoft eclipses the marketplace and restructures it under a single standard.) Viruses and worms exploit holes and in this sense are a good index for oppositional network practices. They propagate through weaknesses in the logical structure of computer code. When an exploit is discovered, the broad homogeneity of computer networks allows the virus to resonate far and wide with relative ease. Networks are, in this sense, a type of massive amplifier for action. Something small can turn into something big very easily.

Anyone who owns a computer or regularly checks e-mail knows that the virus versus antivirus situation changes on a daily basis; it is a game of cloak-and-dagger. New viruses are constantly being written and released, and new patches and fixes are constantly being uploaded to Web sites.

Users must undertake various prophylactic measures ("don't open attachments from unknown senders"). Computer security experts estimate that there are some eighty thousand viruses currently recorded, with approximately two hundred or so in operation at any given moment.[63] Such a condition of rapid change makes identifying and classifying viruses an almost insurmountable task. Much of this changeability has come from developments in the types of viruses, as well. Textbooks on computer viruses often describe several "generations" of malicious code. First-generation viruses spread from machine to machine by an external disk; they are often "add-on" viruses, which rewrite program code, or "boot sector" viruses, which install themselves on the computer's MBR (master boot record) so that, upon restart, the computer would launch from the virus's code and not the

computer's normal MBR. Early antivirus programs performed a calcu-
lation in which the size of program files was routinely checked for
any changes (unlike document files, program files should not change,
and thus a change in the file size indicated an add-on or other type of
virus). Second-generation viruses were able to outmaneuver these
calculations by either ballooning or pruning program code so that it
always remains the same size. Third-generation viruses, such as "stealth"
viruses, went further, being able to intercept and mimic the antivirus
software, thereby performing fake file scans. Fourth-generation viruses
are the opposite of the third generation; they employ "junk code"
and "attack code" to carry out multipronged infiltrations, in effect
overwhelming the computer's antivirus software ("armored" viruses).
However, one antivirus technique has remained nominally effective,
and that is the identification of viruses based on their unique "signature," a string of code that is specific to each virus class. Many antivirus
programs use this approach today, but it also requires a constantly
updated record of the most current viruses and their signatures. Fifth-
generation viruses, or "polymorphic" viruses, integrate aspects of
artificial life and are able to modify themselves *while they replicate and
propagate through networks*. Such viruses contain a section of code—a
"mutation engine"—whose task is to continuously modify its signature code, thereby evading or at least confusing antivirus software.
They are, arguably, examples of artificial life.[64]

*Viruses such as the polymorphic computer viruses are defined by their
ability to replicate their difference. They exploit the network.*

That is, they are able to change themselves at the same time that
they replicate and distribute themselves. In this case, computer viruses
are defined by their ability to change their signature and yet maintain a continuity of operations (e.g., overwriting code, infiltrating as
fake programs, etc.). Viruses are never quite the same. This is, of course,
one of the central and most disturbing aspects of biological viruses—
their ability to continuously and rapidly mutate their genetic codes.
    This ability not only enables a virus to exploit new host organisms
previously unavailable to it but also enables a virus to cross species
boundaries effortlessly, often via an intermediary host organism. There

is, in a way, an "animality" specific to the biological virus, for it acts as a connector between living forms, traversing species, genus, phylum, and kingdom. In the late twentieth century and the early twenty-first, public health organizations such as the WHO and the CDC began to see a new class of diseases emerging, ones that were caused by rapidly mutating microbes and were able to spread across the globe in a matter of days.

These "emerging infectious diseases" are composed of assemblages of living forms: microbe-flea-monkey-human, microbe-chicken-human, microbe-cow-human, or human-microbe-human. In a sense, this is true of all epidemics: in the mid-fourteenth century, the Black Death was an assemblage of bacillus-flea-rat-human, a network of contagion spread in part by merchant ships along trade routes.

Biological viruses are connectors that transgress the classification systems and nomenclatures that we define as the natural world or the life sciences. The effects of this network are, of course, far from desirable. But it would be misleading to attribute maliciousness and intent to a strand of RNA and a protein coating, even though we humans endlessly anthropomorphize the nonhumans we interact with. What, then, is the viral perspective? Perhaps contemporary microbiology can give us a clue, for the study of viruses in the era of the double helix has become almost indistinguishable from an information science. This viral perspective has nothing to do with nature, or animals, or humans; it is solely concerned with operations on a code (in this case, a single-strand RNA sequence) that has two effects—the copying of that code within a host organism, and mutation of that code to gain entry to a host cell.

Replication and cryptography are thus the two activities that define the virus. What counts is not that the host is a "bacterium," an "animal," or a "human." What counts is the code—the number of the animal, or better, the numerology of the animal.

We stress that the viral perspective works through replication *and* cryptography, a conjunction of two procedures. Sticking to our examples of computer and biological viruses, the kind of cryptography involved is predicated on mutation and morphology, on recombining and recalculating as a way of never-being-the-same. The viral perspective is "cryptographic" because it replicates this difference, this paradoxical status of never-being-the-same. Again and again, it is never the same. What astounds us is *not* that a virus is somehow "transgressive," crossing species borders (in the case of biological viruses) or different platforms (in the case of computer viruses). The viral perspective, if indeed we are to comprehend its unhuman quality, is not some rebellious or rogue piece of data infiltrating "the system." What astounds us is that the viral perspective presents the animal being and creaturely life in an *illegible* and *incalculable* manner, a matter of chthonic calculations and occult replications. This is the strange numerology of the animal that makes species boundaries irrelevant.

Given this, it is no surprise that the language and concept of the virus have made their way into computer science, hacking, and information-security discourse. Computer viruses "infect" computer files or programs, they use the files or programs to make more copies of themselves, and in the process they may also employ several methods for evading detection by the user or antivirus programs. This last tactic is noteworthy, for the same thing has both intrigued and frustrated virologists for years. A virus mutates its code faster than vaccines can be developed for it; a game of cloak-and-dagger ensues, and the virus vanishes by the time it is sequenced, having already mutated into another virus. Computer viruses are, of course, written by humans, but the effort to employ techniques from artificial life to "evolve" computer viruses may be another case altogether. The fifth-generation polymorphic viruses are able to mutate their code (thereby eluding the virus signature used by antivirus programs) as they replicate, thus never being quite the same virus.

Viruses are entities that exist solely by virtue of the continual replication of numerical difference.

The virus, in both its biological and computational guises, is an exemplary if ambivalent instance of "becoming" and thus stands in dialogue with Zeno's famous paradoxes. Zeno, like his teacher Parmenides, argues that there cannot be A and not-A at the same time (e.g., if an archer shoots an arrow and measures its distance to a tree by dividing the distance in half each time, the arrow never reaches the tree. But how can the arrow both reach the tree and not reach the tree?). There must be a unity, a One-All behind everything that changes. In a sense, our inability to totally classify biological or computer viruses serves as a counterpoint to this earlier debate. If viruses are in fact defined by their ability to replicate their difference, we may ask, what is it that remains identical throughout all the changes? One reply is that it is the particular structure of change that remains the same—permutations of genetic code or computer code. There is a *becoming-number* specific to viruses, be they biological or computational, a mathematics of combinatorics in which transformation itself—via ever new exploitation of network opportunities—is the identity of the virus.

If the computer virus is a technological phenomenon cloaked in the metaphor of biology, emerging infectious diseases are a biological phenomenon cloaked in the technological paradigm. As with computer viruses, emerging infectious diseases constitute an example of a counterprotocol phenomenon.

In this way, epidemiology has become an appropriate method for studying computer viruses. Emerging infectious diseases depend on, and make use of, the same topological properties that constitute networks. The same thing that gives a network its distributed character, its horizontality, is therefore transformed into a tool for the destruction of the network.

An example is the 2003 identification of SARS (severe acute respiratory syndrome). While there are many, many other diseases that are more crucial to look at from a purely public health perspective, we find SARS interesting as a case study for what it tells us about the unhuman, viral perspective of networks.

The SARS case is noteworthy for the rapidity with which the virus was identified. The World Health Organization estimates that it took a mere seven weeks to identify the virus responsible for SARS (and a mere six days to sequence the virus genome).[65] Compared to the three years it took to identify HIV, and the seven years it took to identify Lyme disease, the SARS case is an exemplary instance of high-technology epidemiology.

In addition, SARS is also an exemplary case of a biological network in action. Between November 2002 (the first SARS cases) and March 2003, an estimated 3,000 cases had been reported in twenty-five countries, with more than 150 deaths worldwide. By June 2003, the number of cases had jumped to more than 8,454, with some 6,793 recoveries and 792 deaths. Countries with the leading number of reported cases include China (with over 5,000), as well as Hong Kong, Canada, Taiwan, Singapore, Vietnam, and the United States.

But SARS is also much more than just a biological network—it brings together other networks such as transportation, institutional, and communications networks (and in ways that often seem to read like a medical thriller novel). In November 2002, the first cases of SARS (then referred to as "atypical pneumonia") appeared in the southern Chinese province of Guangdong. By mid-February 2003, the WHO and other health agencies were alerted to a new type of pneumonia coming out of China. The Chinese government reported some three hundred cases, many in and around Guangdong Province. In late February, a physician who had treated patients with atypical pneumonia in Guangdong returned to his hotel in Hong Kong. The biological network interfaced with the transportation network. The WHO estimates that this physician had, in the process, infected at least twelve other individuals, each of whom then traveled to Vietnam, Canada, and the United States. Days later, Hong Kong physicians reported the first cases of what they begin to call "SARS." A few weeks later, in early March, health care officials in Toronto, Manila, and Singapore reported the first SARS cases. Interfacing institutional and communications networks, the WHO issued a travel advisory via newswire and Internet, encouraging checkpoints in airports for flights to and from locations such as Toronto and Hong Kong. At the same time, the WHO organized an international teleconference among health care administrators and officials (including the CDC), agreeing to share information regarding SARS cases. Uploading of patient data related to SARS to a WHO database began immediately. The professional network interfaced with the institutional network, and further to the computer network. By late March, scientists at the CDC suggested that a mutated coronavirus (which causes the common cold in many mammals) may be linked to SARS. Then, on April 14, scientists at Michael Smith Genome Sciences Centre in Vancouver sequenced the DNA of the SARS coronavirus within six days (to be repeated by CDC scientists a few days later). By April 2003, SARS continued to dominate news headlines, on the cover of *Time*, *Newsweek*, and *U.S. News* concurrently.

While this coordination and cooperation via the use of different networks is noteworthy, on the biological level, SARS continued to transform and affect these same networks. In early April 2003, the U.S. government issued an executive order allowing the quarantine of healthy people suspected of being infected with SARS but who did not yet have symptoms. During March and April 2003, quarantine measures were carried out in Ontario, Hong Kong, Singapore, and Beijing. Residential buildings, hospitals, and public spaces such as supermarkets, cinemas, and shopping malls were all subjected to quarantine and isolation measures. People from Toronto to Beijing were regularly seen wearing surgical masks to ward off infection. By late April, the spread of SARS seemed to stabilize. WHO officials stated that SARS cases peaked in Canada, Singapore, Hong Kong, and Vietnam (though not in China). Many countries reported a decrease in the number of SARS cases, although no vaccine has yet been developed. In late May 2003, U.S. health officials warned that the SARS virus will most likely reappear during the next flu season.

Right away, one should notice something about SARS as an emerging infectious disease. SARS is first of all an example of a biological network. But it is also more than biological.

SARS and other emerging infectious diseases are the new virologies of globalization; the meaning of the term "emerging infectious disease" itself implies this. Emerging infectious diseases are products of globalization. This is because they are highly dependent on one or more networks.

The SARS coronavirus tapped into three types of networks and rolled them into one: (1) the biological network of infection (many times within health care facilities); (2) the transportation network of airports and hotels; and (3) the communications networks of news,

Web sites, databases, and international conference calls. Without a transportation network, the SARS virus may well have been a localized Chinese phenomenon, never reaching as far as Toronto. And while news reports in the United States served mostly to educate a worried public, they also served to heighten the anxiety of a population already sensitized to bioterrorism. Perhaps even more significant is the way in which the SARS virus exploited networks to such an extent that the networks were shut down: buildings quarantined, air travel restrictions enforced, people relocated or isolated.

Emerging infectious diseases such as SARS not only operate across different networks simultaneously but in so doing also transgress a number of boundaries. The lesson here is that network flexibility and robustness are consonant with the transgression of boundaries.

The SARS virus, for instance, crosses the species boundaries when it jumps from animals to humans. It also crosses national boundaries in its travels between China, Canada, the United States, and Southeast Asia. It crosses economic boundaries, affecting the air travel industry, tourism, and entertainment industries, as well as providing initiative and new markets for pharmaceutical corporations. Finally, it crosses the nature–artifice boundary, in that it draws together viruses, organisms, computers, databases, and the development of vaccines. Its tactic is the *flood,* an age-old network antagonism.

But biological networks such as the SARS instance are not just limited to diseases that are said to be naturally occurring. Their network effects can be seen in another kind of biological network—that of bioterrorism. While some may argue that the perceived bioterrorist threat has been more about national security initiatives than about any real threat, what remains noteworthy in the case of bioterrorism is the difficulty in identifying, assessing, and controlling the network that would be created by the strategic release of a harmful biological agent.

Such skirmishes highlight an important point in our understanding of networks: that the networks of emerging infectious diseases and bioterrorism are networks composed of several subnetworks.

In the case of bioterrorism, there is the biological network of the genome of a virus or bacteria and its mode of operating, there is the epidemiological network of infection and spread, there are a number of institutional networks geared toward intervention and prevention (the CDC, the WHO), there is the knowledge network of public education and awareness-raising, there is the "global" pharmaceutical network providing medicines, and there is the political network of mobilizing national security and defense initiatives (the U.S. Health and Human Services' "Operation Bioshield").

The post–September 11 anthrax events are an example of the layering of networks and the diversity of networked control.

On October 25, 2001, following the discovery of the anthrax-tainted letters, House Democrats introduced a $7 billion "bioterrorism bill," which called for increased spending on a national health care surveillance effort, as well as an increase in the national stockpile of vaccines and antibiotics (including the research needed for their development). The day before this, health officials announced that a deal had been struck with Bayer (among the largest transnational "big pharma" corporations) to purchase large amounts of Cipro, the anthrax antibiotic of choice. During this time, several government buildings, including post offices, Senate offices, and businesses, were either temporarily or indefinitely closed off for testing and decontamination.

To the perspective that says the bioterrorist activities involving anthrax were not successful, we can reply that, on the contrary, they have been very successful in generating a state of immanent preparedness on the governmental layer, accompanied by a state of "bio-horror" on the cultural and social layer.[66] This accompanying state of anxiety may or may not have anything to do with the reality of bio-warfare, and that is its primary quality. That anthrax is not a contagious disease like the common cold matters less than the fact that an engineered biological agent has infiltrated components of the social fabric we take for granted—the workplace, the mail system, even subways and city streets. If there is one way in which the bioterrorist anthrax attacks have "targeted" the body, it is in their very proximity to our bodies, which has triggered a heightened cultural and social anxiety about the threat of contagion.

But are we not missing an important distinction in this discussion of SARS and anthrax? Is there not an important distinction between naturally occurring diseases such as SARS and those intentionally constructed and disseminated in bioterrorist acts? International organizations such as the WHO and national ones such as the CDC use the term "emerging infectious disease" for naturally occurring, active diseases that have not yet been observed or identified. A majority of emerging infectious diseases are mutations of common diseases in other species, as is the case with SARS, as well as other emerging infectious diseases (Mad Cow, West Nile, H5N1/bird flu, among others). The description of emerging infectious diseases as "naturally occurring" is meant, among other things, to sharply distinguish them from nonnatural instances of disease—most notably in acts of bioterrorism. This division between emerging infectious diseases and bioterrorism pivots on the question of agency and causality, and the implications of this division are worth considering, if only briefly.

In bioterrorism there is an instrumental use of biology as a weapon—there is a subject, often motivated by ideology, using biology as a weapon. This subject is therefore accountable because there is a clear cause-and-effect relationship. In this case, the subject would then be treated legally as a criminal using a weapon to carry out an act of violence. Yet in the case of emerging infectious diseases, there is no subject. There is, of course, a sometimes hypothetical "patient zero," which serves as the index of a cross-species migration, but there is no agential subject that intentionally "causes" the disease. This attribution of the disease to nature has both a mystifying and alienating effect, especially in the alarmism that much media reportage fosters. There is no subject, no motivation, no original act, and no ideology— only rates of infection, identification of the disease-causing agent, and detection procedures.

From the point of view of causality, emerging infectious diseases and bioterrorism are self-evidently different. Yet from the point of view of their effects—their network effects—they are the same.

That is, while emerging infectious diseases and bioterrorism may have different points of origin, their resultant effects are quite similar—the processes of infection, dissemination, and propagation serve to spread both emerging infectious diseases and bioterrorist agents. For this reason, a number of earlier infectious diseases (such as smallpox or polio) have been the subject of biowarfare research.

The causes may be different, but the results are the same. This should give us pause.

It is not a question of mere numbers such as the number of people infected or sick—and this is not a metaphorical claim, as in the viral "meme" of anthrax. It is a point about different types of biological networks that are also more than biological. The protocol of these networks proceeds by the biological principles of epidemiology (infection rates, death and recovery rates, area of infection, etc.). But their overall organization is akin to the previous claims about computer viruses, for it is precisely the standardization of networks, along with their distributed topology, that enables an emerging infectious disease to maintain a certain level of effectiveness. The air travel industry works exceedingly well, as does the communications media network. Computer viruses and emerging infectious diseases profit from these modern conveniences.

Both emerging infectious diseases and bioterrorism reveal the sometimes uncanny, unsettling, and distinctly nonhuman aspect of networks. There is a defacement of enmity, but an antagonism lingers nonetheless.

A single rogue may send an anthrax-tainted letter, or a single animal may be carrying a lethal virus, but one learns nothing about the network effects—and the network affect—of such diseases by focusing simply on one human subject, one viral sample, one link in the network, or the network's supposed point of origin. Shift from *or* to *and*: it is not emerging infectious diseases *or* bioterrorism, but rather emerging infectious diseases *and* bioterrorism.

Like computer viruses, emerging infectious diseases are frustratingly nonhuman.

True, in extreme cases such as bioterrorism, human agencies are involved in manufacturing, weaponizing, and disseminating lethal diseases. Or in instances of naturally occurring emerging infectious diseases, the decision to share information or to withhold it is regulated by individuals, groups, and institutions. In all these cases, we have legal, governmental, and ethical structures that clearly assign accountability to the persons involved. In other words, we have social structures to accommodate these instances; they fit into a certain paradigm of moral-juridical accountability.

Our contention, however, is that in some ways they fit too neatly. This is precisely the core of the alarmism and anxiety that surround bioterrorism and emerging infectious diseases.

While individuals, groups, or organizations may be responsible for "causing" emerging infectious diseases, it is notoriously difficult to predict the exact consequences of such decisions or to foresee the results of such actions. This is because emerging infectious diseases are not weapons like missiles, planes, or bombs; they are networks, and it is as networks that they function and as networks that they are controlled.

Let us be clear on this point. From the perspective of network logic, the question is not whether there is a difference between computer viruses, bioterrorism, and emerging infectious diseases. On a number of points there are clear and obvious differences, and these differences have been addressed in direct, practical ways. Rather, the question is whether, despite their differences, computer viruses, bioterrorism, and emerging infectious diseases share a common protocol, one that belongs to our current context of the society of control. What prompts us to pose the question in this way is the networked quality of all the previously addressed cases (e-mail worm, SARS, anthrax). It is as if once the biological agent is let loose, the question of agency (and with it, the question of accountability) becomes much more complex. The bioterrorist does not target a person, a group, or even a population. Rather, the bioterrorist exploits a layered and standardized network. In a similar way, a new disease such as SARS does not just repeat its cycle of infection; it mutates, and with its mutations come the opportunistic transgressions of borders of all kinds—

species borders, international borders, and the borders of ethnicity and culture.

We should reiterate that protocological exploits (computer viruses, emerging infectious diseases, and so on) are politically ambivalent in their position within and among networks.

While computer worms piggyback on the global standards of TCP/IP and other Internet protocols, emerging infectious diseases operate on the protocols of biological control, gene expression, and cellular metabolism.

Viruses and diseases are obviously not to be looked at as models for progressive political action. But it is precisely in their ambivalent politics that we see both the plasticity and the fragility of control in networks.

There are as many lessons to be learned from the "failures" of networks as there are from their successes. Perhaps we can note that a network fails only when it works too well, when it provides too little room for change within its grand robustness, as the example of the computer virus illustrates.

In this way, networks fail only when they succeed. Networks cultivate the flood, but the flood is what can take down the network.

There are several lessons to be learned from the ambivalent, non-human examples of computer viruses, emerging infectious diseases, and bioterrorism. To begin with, action and agency (at both the ontological and the political level) will have to be rethought in the context of networks and protocological control. This in turn means reconsidering the relationship between causality and accountability. A single person may be legally accountable for setting loose a computer virus or biological agent, but it is a more complicated manner when one person is accused of all the downstream and indirect consequences of that original action. While the legal system may be good at accountability issues, it is less adept at handling what happens after the original event. Most often, the response is simply to shut down or blockade the network (quarantine, firewalls). Because network effects are almost impossible to predict, network causality is not necessarily the same as network accountability. Especially in cases where networks involve multiple interactions between human subjects or groups, the question of "ethical protocols" comes to the forefront.

In addition, it is in the nature of networks to transgress boundaries of all kinds—institutional, disciplinary, national, technical, and biological. As illustrated previously, a single entity such as a computer worm or pouch of powdered anthrax immediately draws together a range of network nodes (computers, companies, people, software; bodies, hospitals, people, drugs). Network borders exist in a range of ways, including information borders (secure servers), biological borders (inter- and cross-species infection), architectural borders (public spaces, airports, urban environments), and political borders (state and national boundaries). Not only will action need to be reconsidered within networks, but action will need to be reconsidered across networks. If a network contains its own failure, then it also contains its own transgressions of borders.

It is possible to distill these claims into something of a formal description. The following is a definition of the exploit as an abstract machine.

- *Vector:* The exploit requires an organic or inorganic medium in which there exists some form of action or motion.
- *Flaw:* The exploit requires a set of vulnerabilities in a network that allow the vector to be logically accessible. These vulnerabilities are also the network's conditions for realization, its becoming-unhuman.
- *Transgression:* The exploit creates a shift in the ontology of the network, in which the "failure" of the network is in fact a change in its topology (for example, from centralized to distributed).

Counterprotocol

We have derived a few points, then, for instigating political change in and among networks. These might be thought of as a series of challenges for "counterprotocological practice," designed for anyone wishing to instigate progressive change inside biotechnical networks.

First, oppositional practices will have to focus not on a static map of one-to-one relationships but on a dynamic diagram of many-to-many relationships. The diagram must not be anthropomorphic (the gesture, the strike); it must be unhuman (the swarm, the flood).

This is a nearly insurmountable task. These practices will have to attend to many-to-many relationships without making the dangerous mistake of thinking that many-to-many means total or universal. There will be no universals for life. This means that *the counterprotocols of current networks will be pliant and vigorous* where existing protocols are flexible and robust. We're tired of being flexible. Being pliant means something else, something vital and positive. Or perhaps "superpliant" would be a better term, following Deleuze's use of the word in the appendix to his book on Foucault.[67] Counterprotocols will attend to the tensions and contradictions within such systems, such as the contradiction between rigid control implicit in network protocols and the liberal ideologies that underpin them. Counterprotocological practice will not avoid downtime. It will restart often.

The second point is about tactics. In reality, counterprotocological practice is not "counter" anything. Saying that politics is an act of "resistance" was never true, except for the most literal interpretation of conservatism. We must search-and-replace all occurrences of "resistance" with "impulsion" or perhaps "thrust." Thus the concept of resistance in politics should be superseded by the concept of hypertrophy.

Resistance is a Clausewitzian mentality. The strategy of maneuvers instead shows that the best way to beat an enemy is to become a better enemy. One must push through to the other side rather than drag one's heels. There are two directions for political change: resistance implies a desire for stasis or retrograde motion, but hypertrophy is the desire for pushing beyond. The goal is not to destroy technology in some neo-Luddite delusion but to push technology into a hypertrophic state, further than it is meant to go. "There is only one way left to escape the alienation of present-day society: *to retreat ahead of it*," wrote Roland Barthes.[68] We must scale up, not unplug. Then, during the passage of technology into this injured, engorged, and unguarded condition, it will be sculpted anew into something better, something in closer agreement with the real wants and desires of its users.

The third point has to do with structure. Because networks are (technically) predicated on creating possible communications between nodes, oppositional practices will have to focus less on the characteristics of the nodes and more on the quality of the interactions between nodes.

In this sense, the distinction between node and edge will break down. Nodes will be constructed as a by-product of the creation of edges, and edges will be a precondition for the inclusion of nodes in the network. Conveyances are key. From the oppositional perspective, nodes are nothing but dilated or relaxed edges, while edges are constricted, hyperkinetic nodes. Nodes may be composed of clustering edges, while edges may be extended nodes.

Using various protocols as their operational standards, networks tend to combine large masses of different elements under a single umbrella. The fourth point we offer, then, deals with motion: counterprotocol practices can capitalize on the homogeneity found in networks to resonate far and wide with little effort.

Again, the point is not to do away with standards or the process of standardization altogether, for there is no imaginary zone of nonstandardization, no zero place where there is a ghostly, pure flow of only edges. Protocological control works through inherent tensions, and as such, counterprotocol practices can be understood as tactical implementations and intensifications of protocological control.

On a reflective note, we must also acknowledge that networks, protocols, and control are not only our objects of study; they also affect the means and methods by which we perform analysis and critique. Events such as computer viruses or emerging infectious diseases require a means of understanding that draws together a number of disciplines, modes of analysis, and practices. This challenge bears as much on cultural theory and the humanities as it does on computer science, molecular biology, and political theory.

If, as the truism goes, it takes networks to fight networks, then it also takes networks to understand networks, as well.

This is the first step in realizing an ethics and a politics of networks, an activation of a political consciousness that is as capable of the critiquing of protocological control as it is capable of fostering the transformative elements of protocol. What would a network form of praxis be like? Just as network protocols operate not through static relationships, and not by fixed nodes, so must any counterprotocol practice similarly function by new codings, whether in terms of disciplines, methodologies, or practices.[69] In a discussion of intellectuals and power, Deleuze provides a helpful way of further thinking about counterprotocol practices:

> The relationship which holds in the application of a theory is never one of resemblance. Moreover, from the moment a theory moves into its proper domain, it begins to encounter obstacles, walls, and blockages which require its relay by another type of discourse.... Practice is a set of relays from one theoretical point to another, and theory is a relay from one practice to another. No theory can develop without eventually encountering a wall, and practice is necessary for piercing this wall.[70]

Because a network is as much a technical system as it is a political one, any theory addressing networks will have to entertain a willingness to theorize at the technical level.

This not only means a radical interdisciplinarity but also means a willingness to carry theorization, and its mode of experimentation, to the level of protocological practices.

Today to write theory means to write code. There is a powerful exhilaration in the transformation of real material life that guides the writing of counterprotocological code. As Geert Lovink reminds us: "No more vapor theory anymore."[71]

We may speculate, then, that as the instruments of social transformation follow this call to action, the transition from the present day into the future might look something like this:

	Societies of Control . . .	*. . . the Future*
control diagram	cybernetics; protocol	physics; particle swarms
machine	computers	bioinformatics
resistive act	mutation; subversion	desertion; perturbation
delinquent act	randomness	nonexistence
political algorithm	disturbance	hypertrophy
stratagem	security; exception	gaming; inception
historical actor	communities; the people	élan vital; multitude
mode of liberation	neoliberal capitalism	"life-in-common"

PART II
Edges

—Well there it is. There's the common basis for communication.
A new language. An inter-system language.

—But a language only those machines can understand.

—*Colossus: The Forbin Project, 1970*

Edges

The Datum of Cura I

Imagine an art exhibit of computer viruses. How would one curate such a show? Would the exhibition consist of documentation of known viruses, or of viruses roaming live in situ? Would it be more like an archive or more like a zoo? Perhaps the exhibit would require the co-ordination of several museums, each with "honeypot" computers, sacrificial lambs offered up as attractor hosts for the contagion. A network would be required, the sole purpose of which would be to reiterate sequences of infection and replication. Now imagine an exhibit of a different sort: a museum dedicated to epidemics. Again, how would one curate an exhibit of disease? Would it include the actual virulent microbes themselves (in a sort of "microbial menagerie"), in addition to the documentation of epidemics in history? Would the epidemics have to be "historical" to qualify for exhibition? Or would two entirely different types of institutions be required: a museum of the present versus a museum of the past?

In actuality such exhibits already exist. A number of artists have created and shown work using the medium of the computer virus, the most noteworthy being the *Biennale.py* virus, released by the collectives 0100101110101101.org and epidemiC as part of the Venice Biennale in 2001. The work was included in the "I love you" computer virus exhibition curated by Francesca Nori in 2004. Likewise, in the United States, the first museum dedicated to disease was established by the Centers for Disease Control and Prevention (CDC). Called the Global Health Odyssey, it uses the format of the history museum to tell the story of epidemics in history and the CDC's "fight" against those epidemics.

But let us linger for a moment on the biological motifs of both these exhibits, as well as on what it might mean to *curate* them. The act of curating not only refers to the selection, exhibition, and storage of artifacts but also means doing so *with care*, with particular attention to their presentation in an exhibit or catalog. Both "curate" and "curator" derive from the Latin *curare* (to care), a word that is itself closely related to *cura* (cure). Curate, care, cure. At first glance, the act of curating a museum exhibit seems far from the practice of medicine and health care. One deals with culture and history, the other with science and "vital statistics." One is the management of "art," the other the management of "life." But with the act of curating an exhibit of viruses or epidemics, one is forced to "care" for the most misanthropic agents of infection and disease. One must curate that which eludes the cure. Such is the impasse: the best curator would therefore need to be the one who is most "careless." We shall return to this point in a moment.

Today's informatic culture has nevertheless brought together curating and curing in unexpected ways, linked by this notion of *curare*. The very concept of "health care," for instance, has always been bound up with a relation to information, statistics, databases, and numbers (numbers of births, deaths, illnesses, and so forth). Indeed, political economy during the era of Ricardo, Smith, and Malthus implied a direct correlation between the health of the population and the wealth of the nation. Yet public health has also changed a great deal, in part due to advances in technology within the health care industry. There is now talk of "telemedicine," "infomedicine," and "home care." At the most abstract level, one witnesses information networks at play in medical surveillance systems, in which the real-time monitoring of potential public health hazards (be they naturally occurring or the result of an attack) is made possible in a "war-room" scenario.

In these visions of health care—in which the law of large numbers is the content, and network topology is the form—there are also many questions raised. The sociologist Michael Fortun, in his study of population genome projects, wonders if we have moved from classical medicine's care of the body of the patient (what Foucault referred to as a "care of the self") to a more post-Fordist "care of the data," in which the job of public health is increasingly to ensure that the biological bodies of the population correlate to the informatic patterns on the screen.[1]

The "epidemic" exhibits such as *Biennale.py* and the Global Health Odyssey are of interest because they are not simply exhibits that happen to have biological motifs. As different as they are, they put curating and curing into a relationship. It is a relationship mediated by *curare* or care. But what is "care" in this case? It is a type of care that is far from the humanistic and phenomenological notion of person-to-person care; it is a "care of the data" in which the life of information or "vital statistics" plays a central role.

The Datum of Cura II

Return to our imagined exhibitions of viruses and epidemics. What is the temporality specific to the practice of curating? The idea of *curare* (care) in curating and the activity of the curator plays a dual role. One the one hand, the care in curating conceptually tends toward the presentation of the static: collecting, archiving, cataloging, and preserving in a context that is both institutional and architectural. There is a stillness to this (despite the milling about of people in a museum or the awkwardness of an "interactive" exhibit). The care of stillness, within walls, behind glass, is a *historical* stillness. It is a stillness of the past. But there is also always an excess in curating, an opening, however wide or narrow, through which the unexpected happens. As a visitor to an exhibit, one's interpretations and opinions might vary widely from both the curator's original vision and from the interpretations and opinions of other visitors. Or one might not notice them at all, passing over all the care put into curating. Such is the scene: there is either too much ("what's your opinion?") or too little ("I didn't notice").

Curating is not, of course, exclusive to museums and galleries. The motif of curating was common during the Middle Ages, most often in reference to a spiritual guide or pastor who was put in charge of a body of laypeople—people whose souls were in the spiritual care of a "curate." Foucault notes that such a practice entailed a certain form of governing. The dominant biblical metaphor in this case was that of the shepherd and flock. As Foucault's later work shows, this type of caring—a caring-for-others—had its complement in an ethics of care for one's self, a genealogy Foucault locates in classical Greek culture. For the Greeks, the notion of *epimeleia heautou* (care of one-self) not only was an attitude toward self, others, and world but referred to a constant practice of self-observation and self-examination. Central to Foucault's analyses was the fact that this type of care was defined by "actions by which one takes responsibility for onself and by which one changes, transforms, transfigures, and purifies oneself."[2] Here *epimeleia heautou* has as its aim not just the care of the self but the transformation of the self; self-transformation was the logical outcome of self-caring.

However, self-transformation also entails self-destruction. This is a central characteristic of change noted by Aristotle ("coming-to-be" complemented by "passing-away"). Is there a definable point at which self-transformation becomes auto-destruction? The phrase "auto-destruction" was used by Gustav Metzger for many of his performative artworks during the 1960s. In *The Laws of Cool*, Alan Liu describes Metzger's auto-destructive artworks as an early form of what he calls "viral aesthetics." This refers to an aesthetic in which the distinction between production and destruction is often blurred, revealing "a destructivity that attacks knowledge work through technologies and techniques internal to such work."[3] If Metzger is the industrial forerunner of viral aesthetics, then for Liu, the contemporary work of artists like Jodi and Critical Art Ensemble are its heirs. For Liu, such examples of viral aesthetics "introject destructivity within informationalism,"[4] which is so often predicated on the information/noise division.

Curare thus presupposes a certain duplicitous relation to transformation. It enframes, contextualizes, bounds, manages, regulates, and controls. In doing so, it also opens up, unbridles, and undoes the very

control it seeks to establish. It is the point where control and transformation intersect. Which brings us to an ending in the form of a
question: is there a certain "carelessness" to *curare?*

Sovereignty and Biology I

Political thought has long used the body as a metaphor for political
organization. Plato analogizes the political order of the polis with the
biological order of the body and in doing so medicalizes politics. After
having spent the majority of the work discussing the constitution of
a just political order, the *Republic* turns to the forces of dissolution or
decomposition that threaten the body politic. Primary among these is
the descent from concerns of justice to concerns of wealth (oligarchy)
and concerns of appetites (democracy). Though economic health and
basic necessities are central to the proper functioning of the polis, it is
their excess that creates the "illness of a city."[5] For Plato, if oligarchy
represents the excessive rule of wealth for its own sake, then democracy,
in his terms, represents the imbalance between desire and freedom,
in which freedom is always the legitimation for desire. The combination of the two results in the diseased body politic: "When [oligarchy
and democracy] come into being in any regime, they cause trouble, like
phlegm and bile in a body. And it's against them that the good doctor
and lawgiver of a city, no less than a wise beekeeper, must take long-
range precautions, preferably that they not come into being, but if they
do come into being, that they be cut out as quickly as possible, cells
and all."[6] This same logic—a kind of medical sovereignty—is played
out in mechanistic terms in Hobbes's *De Corpore Politico,* and in organicist terms in chapters 13 to 20 of Rousseau's *The Social Contract.*
In the current era of genetics and informatics, has the concept of the
body politic changed? If the understanding of the body changes, does
this also require a change in the understanding of the body politic?

Sovereignty and Biology II

In one of his lectures at the Collège de France, Foucault suggests that
contemporary analyses of power need to develop alternative models to
the tradition of juridical sovereignty: "In short, we have to abandon

the model of *Leviathan*, that model of an artificial man who is at once an automaton, a fabricated man, but also a unitary man who contains all real individuals, whose body is made up of citizens but whose soul is sovereignty."[7] Foucault himself acknowledges the imbrication of sovereignty with the more bottom-up paradigm of discipline. At the same time that disciplinary measures are developed within institutions, a "democratization of sovereignty" takes place, in which the people hold the right to auto-discipline, to accept and in fact demand modes of auto-surveillance in the name of a *biological* security. But the reference that Foucault makes to Hobbes is significant, for it raises a fundamental issue of contemporary political thought: Is it possible to conceive of a body politic without resorting to the paradigm of absolute sovereignty? In other words, can a political collectivity exist without having to transfer its rights to a transcendent body politic?

One of the ways that sovereignty maintains its political power is continually to identify a biological threat. Giorgio Agamben points to the "state of exception" created around what he calls "bare life." Bare life, life itself, the health of the population, the health of the nation—these are the terms of modern biopolitics. By grounding political sovereignty in biology, threats against the biological body politic, in the form of threats against the health of the population, can be leveraged as ammunition for building a stronger sovereign power. Foucault is just as explicit. Medicine, or a medicalization of politics, comes to mediate between the "right of death" and the "power over life": "The development of medicine, the general medicalization of behavior, modes of conduct, discourses, desires, and so on, is taking place on the front where the heterogeneous layers of discipline and sovereignty meet."[8]

Abandoning the Body Politic

There are two states of the body politic. One is the constitutive state, where the body politic is assembled, as Hobbes notes, through "acquisition or institution." This kind of body politic is built on a supposed social contract, or at the least a legitimatized basis of authority, to ensure the "security of life." The other state of the body politic is that of dissolution, the source of fear in virtually every modern political treatise: Machiavelli's plebs or Hobbes's mob rule. Even Locke and Rousseau, who authorize revolution under special conditions when the contract is violated, still express an ambivalence toward this dissolutive state of the body politic. Every political treatise that expresses the first state of the body politic thus also devotes some furtive, discomforting sections to the second. In some cases, this dissolutive body politic is simply chaos, a return to the "state of nature." In other cases, it is a force synonymous with the sovereignty of the people, as it is in Spinoza. Whatever the case, each expression of a constitutive and constituted body politic also posits a dissolutive body politic as its dark side. But there is a problem: the two types of body politic feed into each other through the mechanism of war. We can reiterate Foucault's inversion of Clausewitz: politics is war by other means. Whether the ideal war of the state of nature, or the actual war that continually threatens the civil state, war seems to be the driving force of the two body politics. "In the smallest of its cogs, peace is waging a secret war," wrote Foucault.[9] In this light, perhaps Jean-Luc Nancy's notion of "abandoned being" can be read as a call to abandon the body politic. For Nancy, abandoned being is both the leaving-behind of the being/nonbeing distinction, as well as an assertion of a new fullness, the fullness of desertion: "If from now on being is not, if it has begun to be only its own abandonment, it is because this speaking in multiple ways is abandoned, is in abandonment, and it is abandon (which is also to say openness). It so happens that 'abandon' can evoke 'abundance.'"[10] Abandoning the body politic not only means leaving behind—or deserting—the military foundations of politics but also means a radical opening of the body politic to its own abandon. When the body politic is in abandon, it opens onto notions of the common, the open, the distributed. "What is left is an irremediable scattering, a dissemination of ontological specks."[11]

The Ghost in the Network

Discussing the difference between the living and the nonliving, Aristotle points to the phenomena of self-organized animation and motility as the key aspects of a living thing. For Aristotle the "form-giving Soul" enables inanimate matter to become a living organism.

If life is animation, then animation is driven by a final cause. But the cause is internal to the organism, not imposed from without as with machines. Network science takes up this idea on the mathematical plane, so that geometry is the "soul" of the network. Network science proposes that heterogeneous network phenomena can be understood through the geometry of graph theory, the mathematics of dots and lines. An interesting outcome of this is that seemingly incongruous network phenomena can be grouped according to their similar geometries. For instance, the networks of AIDS, terrorist groups, and the economy can be understood as having in common a particular pattern, a particular set of relations between dots (nodes) and lines (edges). A given topological pattern is what cultivates and sculpts information within networks. To in-form is thus to give shape to matter (via organization or self-organization) through the instantiation of form—a network hylomorphism.

But further, the actualized being of the living network is also defined in political terms. "No central node sits in the middle of the spider web, controlling and monitoring every link and node. There is no single node whose removal could break the web. A scale-free network is a web without a spider."[12] Having-no-spider is an observation about predatory hierarchy, or the supposed lack thereof, and is therefore a deeply political observation. To make this unnerving jump—from math (graph theory) to technology (the Internet) to politics (the acephalous "web without a spider")—politics needs to be seen as following the necessary and "natural" laws of mathematics; that is, networks need to be understood as "an unavoidable consequence of their evolution."[13] In network science, the "unavoidable consequence" of networks often resembles something like neoliberal democracy, but a democracy that naturally emerges according to the "power law" of decentralized networks, themselves to blame for massive planetary inequities. Like so, their fates are twisted together.

Birth of the Algorithm

James Beniger writes that "the idea may have come from late eighteenth-century musical instruments programmed to perform automatically under the control of rolls of punched paper."[14] By 1801

Joseph-Marie Jacquard had developed punch cards to hold encoded mechanical patterns for use in his looms. The art of weaving, allowed some human flexibility as a handicraft, was translated into the hard, coded grammar of algorithmic execution. Then in 1842 Ada Lovelace outlined the first software algorithm, a way to calculate Bernoulli numbers using Charles Babbage's Analytical Engine. The term "algorithm" itself is eponymous of the medieval Persian mathematician Al-Khwarizmi, inventor of the balancing equations and calculations collectively known as algebra. Algorithms always need some processing entity to interpret them—for Jacquard it was the hardware of the loom itself, and for Lovelace it was Babbage's machine. In this sense, algorithms are fundamentally a question of mechanical (or later electronic) processing. Algorithms can deal with contingencies, but in the end they must be finite and articulated in the grammar of the processor so that they may be parsed effectively. Because of this, *the processor's grammar defines the space of possibility for the algorithm's data set.* Likewise, an algorithm is a type of visible articulation of any given processor's machinic grammar.

In 1890 Herman Hollerith used punch cards to parse U.S. census data on personal characteristics. If punch cards are the *mise-en-écriture* (Thomas Levin) of algorithms, their instance of inscription, then in the 1890 census the entire human biomass of the United States was inscribed onto an algorithmic grammar, forever captured as biopolitical data. Today Philip Agre uses the term "grammars of action" to describe the way in which human action is parsed according to specific physical algorithms.[15] Imagine the "noise sequences" that have been erased.

Political Animals

Aristotle's famous formulation of "man as a political animal" takes on new meanings in light of contemporary studies of biological self-organization. For Aristotle, the human being was first a living being, with the additional capacity for political being. In this sense, biology is a prerequisite for politics, just as the human being's animal being serves as the basis for its political being. But not all animals are alike. Deleuze distinguishes three types of animals: domestic pets (Freudian, anthropomorphized Wolf-Man), animals in nature (the isolated species, the lone wolf), and packs (multiplicities). It is this last type of animal—the pack—that provides the most direct counterpoint to Aristotle's formulation. It leads to the following question: if the human being is a political animal, are there also animal politics? Ethnologists and entomologists would think so. The ant colony and insect swarm have long been used in science fiction and horror as the metaphor for the opposite of Western liberal democracies. Even the vocabularies of biology retain the remnants of sovereignty: the queen bee, the drone. What, then, do we make of theories of biocomplexity and swarm intelligence, which suggest that there is no "queen" but only a set of localized interactions that self-organize into a whole swarm or colony? Is the "multitude" a type of animal multiplicity? Such probes seem to suggest that Aristotle based his formulation on the wrong kinds of animals. "You can't be one wolf," of course. "You're always eight or nine, six or seven."[16]

Sovereignty and the State of Emergency

The video game *State of Emergency* offers gamers a chance to be part of an urban riot, a mass melee that has no aim other than to overthrow an anonymous, vaguely named "Corporation." Designed by Rockstar Games in the wake of the 1992 Rodney King riots and the 1999 Battle for Seattle, *State of Emergency* puts artificial life algorithms to good use. One must carefully navigate the chaotic swarm of civilians, protesters, and riot police. The game has no aim except to incite riot, and it is unclear whether the titular "state of emergency" refers to the oppressive corporate State or the apparent chaos that ensues. In other words: is the State of emergency also a state of emergency? Except for military simulation games, rarely do games so explicitly make politics part of their gameplay. One can imagine the game played from the other side—that of the riot police. Here the goal would be crowd control, surveillance, and military blockading. The computer skills necessary for playing either scenario amount to network management tasks. Either you are infiltrating the city and destabilizing key nodes, or you are fortifying such nodes. The lesson of *State of Emergency* is not that it promotes an anarchic ideology but that, in the guise of anarchic ideology, it promotes computer and network management skills. Again following Agamben, modern sovereignty is based not on the right to impose laws but on the ability to suspend the law, to claim a state of emergency. In a way, *State of Emergency* is sovereignty through the back door: inside the screen-based rioting, what is at play is the new sovereignty of networks, control, and the fetish of information.

In this sense, forms of informatic play should be interrogated not as a liberation from the rigid constraints of systems of exchange and production but as the very pillars that prop those systems up. The more video games appear on the surface to emancipate the player, raising his or her status as an active participant in the aesthetic moment, the more they enfold the player into codified and routinized models of behavior. Only eight buttons (mirrored in eight bits) are available for the entire spectrum of expressive articulation using the controller on the Nintendo Entertainment System. A PlayStation running *State of Emergency* supplements this with a few more channels of codified input. Just as the school, in Foucault, was merely preschool for the learned behavior necessary for a laboring life on the factory floor, so games from *State of Emergency* to *Dope Wars* are training tools for life inside the protocological network, where flexibility, systemic problem solving, quick reflexes, and indeed play itself are as highly valued and commodified as sitting still and hushing up were for the disciplinary societies of modernity.

Fork Bomb I[17]

```
#!/usr/bin/perl
while (print fork," ") {
  exit if int rand(1.01);
}
```

Epidemic and Endemic

One of the results of the American-led war on terror has been the increasing implosion of the differences between emerging infectious diseases and bioterrorism. Not so long ago, a distinction was made between emerging infectious disease and bioterrorism based on their cause: one was naturally occurring, and the other resulted from direct human intervention. International bodies such as the World Health Organization and the United Nations still maintain this distinction, though only vaguely. The U.S. government, in the meantime, has long since dispensed with such niceties and as a result has radically streamlined the connections between the military, medicine, and the economics of drug development. A White House press release states: "In his 2003 Budget, the President has proposed $1.6 billion to assist State and local health care systems in improving their ability to manage both contagious and non-contagious biological attacks." Similarly, a 2003 press release describes Project BioShield as a "comprehensive effort to develop and make available modern, effective drugs and vaccines to protect against attack by biological and chemical weapons or other dangerous pathogens."

The implication of the word "or"—biological weapons *or* other pathogens—signals a new, inclusive stage in modern biopolitics. Regardless of the specific context, be it disease or terrorist, the aim is to develop a complete military-medical system for "alert and response" to biological threats. Context and cause are less important than the common denominator of biological effect. It matters little whether the context is terrorism, unsafe foods, compromised federal regulation of new drugs, or new virus strains transported by air travel. What matters is that what is at stake, what is always at stake, is the integrity of "life itself."

This U.S. program of military, medical, and pharmaceutical governance ushers in a politics of "medical security." Medical security aims to protect the population, defined as a biological and genetic entity, from any possible biological threat, be it conventional war or death itself. What this also means is that the biological threat—the inverse of any biological security—is a permanent threat, even an existential threat. It is a biological angst over "death itself" (the biopolitical inverse of "life itself"). This requires a paradigm in which "the population" can be regarded as simultaneously biological and political. As Foucault notes, "At the end of the eighteenth century, it was not epidemics that were the issue, but something else—what might broadly be called endemics, or in other words, the form, nature, ex-

tension, duration, and intensity of the illnesses prevalent in a population.... Death was now something permanent, something that slips
into life, perpetually gnaws at it, diminishes it and weakens it."[18] It is
clear that, in this context, there is no end to biological security; its
job is never finished and, by definition, can never be finished. If there
is one site in which the state of emergency becomes the norm, it is
this site of nondistinction between war and disease, terrorism and
endemic.

Network Being

Information networks are often described as more than mere tools or
relations, but rather as a "global village" or a "collective consciousness." What is it about networks that impels us to describe them as
somehow being alive? For Heidegger, however, the question of *being*
and the question of *being alive* are two different things. Too often an
inquiry of being alive presupposes the self-evident existence of being.
The fields of anthropology, psychology, and biology begin their analyses on the question of "life itself," its modalities and characteristics,
its laws and behaviors, its properties and taxonomies. Rarely do they
ever inquire into the existence as such of "life itself," the almost con-
frontational factuality of the *being* of life (what Levinas described as
the impersonal "horror of the 'there is'...."). As Heidegger notes, "In
the question of the being of human being, this cannot be summarily
calculated in terms of the kinds of being of body, soul, and spirit which
have yet first to be defined."[19]

But it is precisely this question that the sciences of life jump over,
in favor of exclusively anthropomorphic inquiries of psychology or
biology. "Life itself" is always questioned, but the existence *as such* of
life, the being of life, is not regarded as a problem. Such knowledge—
as in the life sciences—thus continues with an assumption of having
understood the very existence as such of living beings. One begins
with Darwinian evolution, with developmental genetics, with studies
of biological morphogenesis, with the genetic factors in health and
disease. "The ontology of life takes place by way of a privative interpretation. It determines what must be the case if there can be anything like just-being-alive."[20] The *life* sciences thus become, in this

regard, reduced to the *human* sciences. For Heidegger, this absence shows itself as a "missing ontological foundation," on which "life itself," and specifically human life, is understood, without recourse to the always-mystical and unstated "being of life" on which it is based.

Our questions are: At what point does the difference between "being" and "life" implode? What would be the conditions for the nondistinction between "being" and "life"? Perhaps this is where the life sciences get hung up. They are confronted with anomalies, anomalies that cross species barriers, that are at once "faceless" and yet "living": single-celled organisms known as myxomycetes (such as the *Physarum* or *Dictostylium*), which, during their life cycles, may be either an amoeba, a motile cell with flagellum, or a plantlike structure giving off spores. Or the famous limit case of the virus. Is it alive? It contains genetic material and is able to reproduce (or at least to replicate). It shows a high degree of genetic adaptability in its mutations and its ability to cross species boundaries. But it is not much more than a strand of RNA and a protein coating. Then, on the opposite side of the scale, there is the infamous case of Gaia . . .

What Heidegger's point makes clear is that the question of "life" has traditionally been separate from, but dependent on, an unquestioned notion of "being." In a way, the example of network science presents us with the opposite case: a concept of "being" is arrived at by a privative definition of "life." Network science, it would seem, assumes a minimally vitalistic aspect of networks, an assumption that informs its studies of networks of all types, networks that all share a being common to networks: "Whatever the identity and the nature of the nodes and links, for a mathematician they form the same animal: a graph or a network."[21] Network science's reliance on universality, ubiquity, and a mathematical model suggests that it is really a *metaphysics of networks*. It seeks a universal pattern that exists above and beyond the particulars of any given network. For this reason, network science can study AIDS, terrorism, and the Internet all as the same kind of being—a network.

The philosophical impact of this view is that of *network being*, a *Dasein* specific to network phenomena. However, what it means specifically is confused. Does it mean the experience of being (in) a network, a new network phenomenology? Does it mean the existence of abstract, mathematical properties across different networks? Networks are said to have a "life of their own," but we search in vain for the "life" that is specific to networks, *except their being as networks*. On the one hand, the proof of the existence as such of living organisms is their living. On the other hand, the proof of the living aspects of networks is their existence as such, that is, their being. The question of "life" and the question of "being" seem always to imply each other, but never to meet.

Good Viruses (SimSARS I)

As the network paradigm gains momentum—in biology, in communications, in international politics—we may be seeing a set of new, network-based strategies being developed on all levels. In some instances, this will be a welcome change. In others, we may not even have the conceptual language to describe the sorts of changes taking place.

One example of this is the increasing attempt to respond to network threats with network solutions. The prevalence of computer viruses, worms, and other bits of contagious code has prompted a new paradigm in computer security: an automated, network-based "vaccine" code that would, like the viruses and worms themselves, circulate autonomously through the networks, detecting flaws and vulnerabilities and automatically fixing them. With tens of thousands of viruses cataloged to date, and a few hundred active at any given moment, there is a secret war being waged behind the tranquillity of the screen.

The good-virus concept is not limited to the digital domain, however. Epidemiologists have long understood that infectious diseases take advantage of a range of networks, many of them human made: biological networks of humans and animals, transportation networks, communications networks, media networks, and sociocultural networks. Media and sociocultural networks can work as much in favor of the virus as against it—witness the pervasive media hype that surrounds any public health news concerning emerging infectious diseases. This is why public health agencies pay particular attention to the use of communications and media networks—when used effectively, they can serve as an awareness and preventive measure. But when does a worst-case scenario turn into alarmism or hype, the stuff of medical thriller movies?

This is so much more the case when emerging infectious diseases are paired alongside the war on terrorism. In many ways, the pairing of emerging infectious disease and bioterrorism is something that is programmatically supported by the U.S. government. While bioterrorism has certainly existed for some time, it is hard to dismiss the heightened anxieties surrounding any news item concerning an infectious disease. The question up front is always: is it a bioterrorist attack? The unspoken thought, the really frustrating thought, is that the most effective bioterrorist attack may be nearly indistinguishable from a naturally occurring disease. Officials from the U.S. Department of Health and Human Services have in the past likened their efforts to the military agenda in Iraq, and public health policy in the United States after September 11, 2001, often combines medical and military research.[22] The disease-as-war metaphor is not new, but it takes on a new guise in the era of networks. If, as we are told, we are fighting "a new kind of war" based on networks, and if "war" has historically been the most common metaphor for talking about disease, then are we also fighting a new kind of medical terror, a new kind of biopolitical war?

What would be the medical analogy, then, for counterterrorist operations and counterinsurgency units? Consider the "good-virus" model applied to the outbreak of an emerging infectious disease: An epidemic is identified, and owing to its networked nature, a counter-network deploys to confront it. An engineered microbe containing a vaccine to the epidemic agent is then released (via aerosol drones) into infected "hot zones," and the microbial netwar is allowed to run its course. Paradoxically, the good virus will succeed in administering the vaccine only if its rate of infection surpasses that of the bad virus. This nexus of disease, medicine delivery, and military logistics is what we can expect in future evolutions of warfare.

Medical Surveillance (SimSARS II)

The condensed, almost aphoristic quality of many recent epidemics continually serves to remind us of the *intensive* nature of networks.

Consider current developments in the practice of medical surveillance. The goal of the "syndromic surveillance" systems being developed by the Centers for Disease Control and Prevention (CDC) within the United States is to implement a real-time, nationwide system for detecting significant anomalies in public health data—anomalies such as the clustering of specific symptoms in a particular city that could point to a possible outbreak or bioterrorist attack.[23] With such systems, we see the layering of different networks on top of each other. Communication is deployed to combat contagion. Distribution of patient records, travel advisories, distribution and delivery of vaccines, identification of the disease-causing agent, selected quarantine: *an information network is used to combat a biological network*. This is the case internationally, as well. The WHO's Global Outbreak Alert and Response Network has as one of its primary aims the insurance that "outbreaks of potential international importance are rapidly verified and information is quickly shared within the Network." These efforts aim to use information networks as key communication tools in times of crisis—be they naturally occurring or intentionally caused—and to that extent are meeting the same challenge given to the original designers of the Internet itself. For pioneering network engineers like Paul Baran, the crisis was the Cold War nuclear threat. For the CDC, it is current biological threats or, rather, the threat of biology. The looming doomsday of the nuclear disaster has been replaced by the ongoing crisis of the biological disaster.

The mere existence of medical surveillance is not problematic in itself—in many cases, it can serve as a benefit for public health. The key issue lies in the relationship between disease, code, and war. Military battles are becoming increasingly virtual, with a panoply of computer-based and information-driven weaponry. And the idea of disease as war has a long history. However, it is foreseeable that the issue of what constitutes "health data" may become a point of some controversy. Concerns over public health will be the Trojan horse for a new era of increased medical surveillance and with it a new militarization of medicine. The institutions of medical surveillance will be almost indistinguishable from national security initiatives and will have shared goals and techniques. While the WHO uses medical data from patients from around the world, it is foreseeable that health data may soon be required in advance from both infected and non-infected organisms. (We are already witnessing this in the areas of genetic screening, genetic counseling, and DNA fingerprinting.)

Imagine a computer simulation like the popular game *SimCity* in which the player develops, builds, and manages a city. But imagine that instead of managing a whole city, the goal is to manage the medical health of the city's inhabitants; instead of being the mayor of SimCity, the player is an official CDC virus hunter. The goal of the game is to watch for potential disease outbreaks and also manage the health of the population on a regular basis (including hospital funding, emergency services, and research centers). This would help illustrate the future of the network model of public health, itself already fully digitized, online, and multiplayer.

In the informatic mode, *disease is always virtual*. It creates a virtual state of permanent emergency wherein infection is always kept just out of reach. And the state of permanent emergency can only be propped up by means of better and better medical surveillance systems.

Feedback versus Interaction I

In the twentieth century there came to pass an evolution in the nature of two-way communication within mass media. This evolution is typified by two models: feedback and interaction. The first model consists of what Beniger calls the mass feedback technologies:

> Market research (the idea first appeared as "commercial research" in 1911), including questionnaire surveys of magazine readership, the Audit Bureau of Circulation (1914), house-to-house interviewing (1916), attitudinal and opinion surveys (a U.S. bibliography lists nearly three thousand by 1928), a Census of Distribution (1929), large-scale statistical sampling theory (1930), indices of retail sales (1933), A. C. Nielsen's audimeter monitoring of broadcast audiences (1935), and statistical-sample surveys like the Gallup Poll (1936).[24]

These technologies establish two-way communications; however, like the media they hope to analyze, the communication loop here is not symmetrical. Information flows in one direction, from the viewing public to the institutions of monitoring.

Contrast this with the entirely different technique of two-way communication called interaction. As a technology, interaction does not simply mean symmetrical communication between two parties. Instead we use interaction to mean an entire system of communicative peers, what Paul Baran called a "distributed network" of communication. We can offer here a list of *interactive* communications technologies to complement Beniger's just-cited list of feedback technologies:

- Paul Baran's description of distributed communications (1964)
- recombinant DNA and the practice of gene-splicing (1973)
- the ARPANET's mandatory rollover to the TCP/IP protocol suite (1983)
- emerging infectious diseases (1980–2000)
- the Gnutella protocol (2000)

Thus interaction happens in an informatic medium whenever there exists a broad network of communicative pairs or multiples, and in which each communicative peer is able to physically affect the other. This ostensibly does not happen in mass media like cinema or television because the audience is structurally unable to achieve a symmetrical relationship of communication with the apparatus (no matter how loudly one yells back at the screen). Interaction happens in the technology of gene-splicing because both sides are able to physically change the system: the scientist changes the physical system by inserting a genetic sequence, while DNA is the informatic code that teleonomically governs the development of physical life. Interaction happens in the Internet protocols for the same reason: protocols interact with each other by physically altering and prepending lesser protocological globs.

Feedback versus Interaction II

As models for two-way communication, feedback and interaction also correspond to two different models of control. Feedback corresponds to the cybernetic model of control, where despite communication occurring bidirectionally between two parties, one party is always the controlling party and the other the controlled party. A homeostatic machine controls the state of a system, not the reverse. Mass media like television and radio follow this model. Interaction, on the other hand, corresponds to a networked model of control, where decision making proceeds multilaterally and simultaneously.

Many today say that new media technologies are ushering in a new era of enhanced freedom and that technologies of control are waning. We say, on the contrary, that *double the communication leads to double the control*. Since interactive technologies such as the Internet are based on multidirectional rather than unidirectional command and control, we expect to see an exponential increase in the potential for exploitation and control through techniques such as monitoring, surveillance, biometrics, and gene therapy. At least the unidirectional media of the past were ignoring half the loop. At least television did not know if the home audience was watching or not. As the mathematicians might say, television is a "directed" or unidirectional graph. Today's media have closed the loop; they are "undirected" or bidirectional graphs. Today's media physically require the maintained, constant, continuous interaction of users. This is the political tragedy of interactivity. We are "treading water in the pool of liquid power," as Critical Art Ensemble once put it.[25]

We long not for the reestablishment of lost traditions of solidification and naturalization as seen in patriarchy or conservatism. We long for the opposite memory: the past as *less* repressive from the perspective of informatic media. Television was a huge megaphone. The Internet is a high-bandwidth security camera. We are nostalgic, then, for a time *when organisms didn't need to produce quantitative data about themselves*, for a time when one didn't need to report back.

Julian Stallabrass: "There is a shadowy ambition behind the concept of the virtual world—to have everyone safely confined to their homes, hooked up to sensory feedback devices in an enclosing, interactive environment which will be a far more powerful tool of social control than television."[26]

Or Vilém Flusser: "An omnipresent dialogue is just as dangerous as an omnipresent discourse."[27]

Rhetorics of Freedom

While tactically valuable in the fight against proprietary software, open source is ultimately flawed as a political program. Open source focuses on code in isolation. It fetishizes all the wrong things: language, originality, source, the past, stasis. To focus on inert, isolated code is to ignore code in its context, in its social relation, in its real experience, or actual dynamic relations with other code and other machines. Debugging never happens through reading the source code, only through running the program. Better than open source would be *open runtime*, which would prize all the opposites: open articulation, open iterability, open practice, open becoming.

But the notion of open runtime may also mislead, given its basis in rhetoric of the relative openness and closedness of technological systems. The rhetoric goes something like this: Technological systems can be either closed or open. Closed systems are generally created by either commercial or state interests—courts regulate technology, companies control their proprietary technologies in the marketplace, and so on. Open systems, on the other hand, are generally associated with the public and with freedom and political transparency. Geert Lovink contrasts "closed systems based on profit through control and scarcity" with "open, innovative standards situated in the public domain."[28] Later, in his elucidation of Castells, Lovink writes of the opposite, a "freedom hardwired, into code."[29] This gets to the heart of the freedom rhetoric. If it's hardwired, is it still freedom? Instead of guaranteeing freedom, the act of hardwiring suggests a limitation on freedom. And in fact that is precisely the case on the Internet, where strict universal standards of communication have been rolled out more widely and more quickly than in any other medium throughout history. Lessig and many others rely heavily on this rhetoric of freedom.

We suggest that this opposition between closed and open is flawed. It unwittingly perpetuates one of today's most insidious political myths, that the state and capital are the two sole instigators of control. *Instead of the open/closed opposition, we suggest an examination of the alternative logics of control.* The so-called open control logics, those associated with (nonproprietary) computer code or with the Internet protocols, operate primarily using an informatic—or, if you like, material—model of control. For example, protocols interact with each other by physically altering and amending lower protocological objects (IP prefixes its header onto a TCP data object, which prefixes its header onto an http object, and so on). But on the other hand, the so-called closed logics of state and commercial control operate primarily using a social model of control. For example, Microsoft's commercial prowess is renewed via the social activity of market exchange. Or digital rights management (DRM) licenses establish a social relationship between producers and consumers, a social relationship backed up by specific legal realities (such as the 1998 Digital Millennium Copyright Act).

From this perspective, we find it self-evident that informatic (or material) control is equally powerful as, if not more so than, social control. If the topic at hand is one of control, then the monikers of "open" and "closed" simply further confuse the issue. Instead we would like to speak in terms of alternatives of control whereby the controlling logic of both "open" and "closed" systems is brought out into the light of day.

A Google Search for My Body

The expectation is that one is either online or not. There is little room for *kind of* online or *sort of* online. Network status doesn't allow for technical ambiguity, only a selection box of discrete states. It is frustrating, ambiguity is, especially from a technical point of view. It works or it doesn't, and when it doesn't, it should be debugged or replaced. To be online in a chronically ambiguous state is maddening, both for those communicating and for the service provider. The advent of broadband connectivity only exacerbates the problem, as expectations for uninterrupted uptime become more and more inflexible. One way to fix the ambiguity is to be "always on," even when asleep, in the bathroom, or unconscious. All the official discourses of the Web demand that one is either online and accounted for, or offline and still accounted for. (This is the idealistic ubiquity of wireless connectivity—the very air you breathe is a domain of access, harkening back to radio's domain of dead voices on air.) Search engines are the best indicator of this demand. Bots run day and night, a swarm of surveillance drones, calling roll in every hidden corner of the Web. All are accounted for, even those who record few user hits. Even as the Web disappears, the networks still multiply (text messaging, multiplayer online games, and so on). The body becomes a medium of perpetual locatability, a roving panoply of tissues, organs, and cells orbited by personal network devices.

Divine Metabolism

Despite, or because of, the popular notion of information as immaterial, information constantly relates to *life-forms*. Life-forms are not merely biological but envelop social, cultural, and political forms as well. Life-forms are the nondistinction between these. Life-forms posit the polyvalent aspect of life, all the while positing something, however inessential, called "life." The foundation of no foundation. Life has many aspects (social, cultural, economic, genetic), and not all of those aspects have an equal claim on life—that is the attitude of the life-form.

But life-forms are also the opposite: the production of a notion of "life itself," a notion of life-forms that is unmediated, fully present, and physical. This notion of the "thing itself" acts as the foundation of life-forms, the point beyond which "life itself" cannot be more immediate. Paradoxically, this is precisely the point at which the more-than-biological must enter the frame. Life-forms are similar to what Marx called the "inorganic body": "Nature is man's *inorganic body*— nature, that is, insofar as it is not itself the human body. Man lives on nature—means that nature is his *body*, with which he must remain in continuous intercourse if he is not to die."[30] But despite his acute analyses, Marx is ambiguous over whether the inorganic body is something other than non- or preindustrial society. Even if we take the inorganic body broadly as "environment," we are still left with the contradictory separation between individual and environment.

Nevertheless, this is the nascent biopolitical aspect that Marx left unexplored—the relation between metabolism and capitalism, between what he called "social metabolism" and political economy. We leave it to Nietzsche to respond: "The human body, in which the most distant and most recent past of all organic development again becomes living and corporeal, through which and over and beyond which a tremendous inaudible stream seems to flow: the body is a more astonishing idea than the old 'soul.' . . . It has never occurred to anyone to regard his stomach as a strange or, say, a divine stomach."[31]

Perhaps networks are the site in which life-forms are continually related to control, where control works through this continual relation to life-forms.

Fork Bomb II

```
#!/usr/bin/perl
while(1){
  if($x = not fork){
    print $x;
  } else {
    print " ";
  }
  exit if int rand(1.03);
}
```

The Paranormal and the Pathological I

In his book *The Normal and the Pathological*, Georges Canguilhem illustrates how conceptions of health and illness historically changed during the eighteenth and nineteenth centuries. Central to Canguilhem's analyses is the concept of "the norm" (and its attendant concepts, normality and normativity), which tends to play two contradictory roles. On the one hand, the norm is the average, that which a statistically significant sector of the population exhibits—a kind of "majority rules" of medicine. On the other hand, the norm is the ideal, that which the body, the organism, or the patient strives for but may never completely achieve—an optimization of health. Canguilhem notes a shift from a quantitative conception of disease to a qualitative one. The quantitative concept of disease (represented by the work of Broussais and Bernard in physiology) states that illness is a deviation from a normal state of balance. Biology is thus a spectrum of identifiable states of balance or imbalance. An excess or deficiency of heat, "sensitivity," or "irritability" can lead to illness, and thus the role of medicine is to restore balance. By contrast, a qualitative concept of illness (represented by Leriche's medical research) suggests that disease is a qualitatively different state than health, a different mode of biological being altogether. The experience of disease involving pain, fevers, and nausea is an indicator of a wholly different mode of biological being, not simply a greater or lesser state of balance. In this case, medicine's role is to treat the symptoms as the disease itself.

However, it is the third and last transition in concepts of illness that is the most telling—what Canguilhem calls "disease as error." Molecular genetics and biochemistry configure disease as an error in the genetic code, an error in the function of the program of the organism. This is the current hope behind research into the genetic mechanisms of a range of diseases and disorders from diabetes to cancer. But what this requires is another kind of medical hermeneutics, one very different from the patient's testimony of the Hippocratic and Galenic traditions, and one very different from the semiotic approach of eighteenth-century pathological anatomy (where lesions on the tissues are signs or traces of disease). The kind of medical hermeneutics required is more akin to a kind of occult cryptography, a deciphering of secret messages in genetic codes. *Disease expresses itself not via the patient's testimony, not via the signs of the body's surfaces, but via a code that is a kind of key or cipher.* The hope is that the *TP53* gene is a cipher to the occulted book of cancerous metastases, and so on. The "disease itself" is everywhere and nowhere—it is clearly immanent to the organism, the body, the patient, but precisely because of this immanence, it cannot be located, localized, or contained (and certainly not in single genes that "cause" disease). Instead disease is an informatic expression, both immanent and manifest, that must be mapped and decoded.

The Paranormal and the Pathological II

One habitually associates epidemics with disease, and disease with death. And for good reason; experience often shows the link between disease and death, or at least between disease and a falling-away from health. But death is not a disease, and disease does not inevitably lead to death. Nevertheless, a historical look at the representations of epidemics suggests that an aura of death—but death as supernatural—surrounds epidemics in both their medical and nonmedical contexts.

In his study of the modern clinic and the "medical gaze," Michel Foucault points to the shift in perspective inaugurated by Bichat in the nineteenth century. Whereas classical medicine largely viewed disease as external to the body, Bichat's notion of a "pathological anatomy" suggested that disease was not separate from the body but rather derived within life itself. Bichat's emphasis on autopsy intended to show that disease was a kind of vital process in the body, but a process that could be differentiated from the effects of death (rigor mortis, decay). Such an approach would lead to the patho-physiological study of fevers and other maladies not as universal, external entities but as dynamic, vital processes within the body. Thus "the idea of a disease attacking life must be replaced by the much denser notion of pathological life."[32]

But the work of Bichat, Broussais, and others focused primarily on the individual body of the patient or the cadaver. What about mass bodies, disease as a mass phenomenon? What about diseases whose etiology includes their modes of transmission and contagion? What of a "pathological life" that exists *between* bodies? Such questions ask us to consider disease in the form of epidemics, contagion, transmission—that is, as networks. Any instance of epidemics poses what Foucault called "the problem of multiplicities": if the processes of contagion, transmission, and distribution have no "center" and are multicausal, then how can they be prevented (or even preempted)? In short, epidemics are not simply medical situations; they are also political ones. If epidemics are networks, then the problem of multiplicities in networks is the tension between sovereignty and control.

Certainly there have been numerous approaches to this problem, and the history of public health provides us with politicized examples of enforced quarantines, ad hoc public health committees, the keeping of "death tables," and the implementation of vaccination programs. But there is another side to this question, another kind of sovereignty, one that medical histories often gloss over. This other kind of sovereignty is located not in governments but in the domain of the supernatural, a divine or demonic sovereignty. Popular interpretations of epidemics throughout history often make appeals to the supernatural: the plague is a sign of divine retribution (for the colonized), a sign of divine providence (for the colonizer), a harbinger of the apocalypse, a punishment of the hubris of humanity, even mystified in modern times as the "revenge of nature." Such representations are not limited to biological epidemics; in the network society, they are also found in informational-biological hybrids: the "metrophage," the "gray goo" problem of nanotech, the "infocalypse," and so on. Such narratives and representations can be seen as attempts to recentralize the question of sovereignty in networks. But in this case, sovereignty is scaled up to the level of the divine or demonic, an agency that may be identified but remains unknowable and decidedly nonhuman. It is as if to say, "there is no one in control, except at an order we cannot fathom." A question of theology, to be sure.

Universals of Identification

Request for Comments (RFC) number 793 states one of the most fundamental principles of networking: "Be conservative in what you do, be liberal in what you accept from others." As a political program, this means that communications protocols are *technologies of conservative absorption*. They are algorithms for translating the liberal into the conservative. And today the world's adoption of universal communications protocols is nearing completion, just as the rigid austerity measures of neoliberal capitalism have absorbed all global markets.

Armand Mattelart wrote that the modern era was the era of universal standards of communication. The current century will be the era of universal standards of identification. In the same way that universals of communication were levied to solve crises in global command and control, the future's universals of identification will solve today's crises of locatability and identification. The problem of the criminal complement is that *they can't be found*. "To know them is to eliminate them," says the counterinsurgency leader in *The Battle of Algiers*. The invention of universals of identification, the ability to locate physically and identify all things at all times, will solve that problem. In

criminal cases, psychological profiling has given way to DNA match-
ing. In consumer products, commodity logistics have given way to
RFID databases. Genomics are the universal identification of life in
the abstract; biometrics are the universal identification of life in the
particular; collaborative filters are the universal identification of life
in the relational.

The twentieth century will be remembered as the last time there
existed nonmedia. In the future there will be a coincidence between
happening and storage. After universal standards of identification are
agreed on, real-time tracking technologies will increase exponentially,
such that almost any space will be iteratively archived over time using
Agre's "grammars of action." Space will become rewindable, fully simu-
lated at all available time codes. Henceforth the lived environment
will be divided into identifiable zones and nonidentifiable zones, and
the nonidentifiables will be the shadowy new criminal classes.

RFC001b: BmTP

A technological infrastructure for enabling an authentic integration
of biological and informatic networks already exists. In separate steps,
it occurs daily in molecular biology labs. The technologies of genomics enable the automation of the sequencing of DNA from any
biological sample, from blood, to test-tube DNA, to a computer file of
text sequence, to an online genome database. And conversely, researchers regularly access databases such as GenBank for their research
on in vitro molecules, enabling them to synthesize DNA sequences
for further research. In other words, there already exists, in many
standard molecular biology labs, the technology for encoding, recod-
ing, and decoding biological information. From DNA in a test tube
to an online database, and back into a test tube. In vivo, in vitro, *in
silico*. What enables such passages is the particular character of the
networks stitching those cells, enzymes, and DNA sequences together.
At least two networks are in play here: the informatic network of the
Internet, which enables uploading and downloading of biological information and brings together databases, search engines, and specialized hardware. Then there is the biological network of gene expression that occurs in between DNA and a panoply of regulatory

proteins, processes that commonly occur in the living cell. The current status of molecular biology labs enables the layering of one network onto the other, so that the biological network of gene expression, for instance, might literally be mapped onto the informatic network of the Internet. The aim would thus be to "stretch" a cell across the Internet. At location A, a DNA sample in a test tube would be encoded using a genome sequencing computer. A network utility would then take the digital file containing the DNA sequence and upload it to a server (or relay it via a peer-to-peer application). A similar utility would receive that file and then download it at location B, from which an oligonucleotide synthesizer (a DNA synthesis machine) would produce the DNA sequence in a test tube. On the one hand, this would be a kind of molecular teleportation, requiring specialized protocols (and RFCs), not FTP, not http, but BmTP, a *biomolecular transport protocol.* Any node on the BmTP network would require three technologies: a sequencing computer for encoding (analog to digital), software for network routing (digital to digital), and a DNA synthesizer for decoding (digital to analog). If this is feasible, then it would effectively demonstrate the degree to which a single informatic paradigm covers what used to be the mutually exclusive domains of the material and the immaterial, the biological and the informatic, the organism and its milieu.

Fork Bomb III

```
#!/usr/bin/perl
while (push(@X,fork) && (rand(1.1)<1)) {
  for(@X) {
    ($_ > $$) ? print "O"x@X : print "o"x@X;
  }
}
```

Unknown Unknowns

Fredric Jameson wrote: it is easier to imagine the deterioration of the earth and of nature than the end of capitalism. The nonbeing of the present moment is by far the hardest thing to imagine. How could things have been otherwise? What is it—can one ever claim with certainty—that hasn't happened, and how could it ever be achieved? "Reports that say that something hasn't happened are always interesting to me," Secretary of Defense Donald Rumsfeld said on the morning of February 12, 2002, responding to questions from the press about the lack of evidence connecting Iraqi weapons of mass destruction with terrorists. "Because as we know, there are known knowns; there are things we know we know. We also know there are known unknowns; that is to say we know there are some things we do not know. But there are also unknown unknowns—the ones we don't know we don't know." There is the unknown soldier. But this is a known unknown, a statistical process of elimination. It is the unknown unknown that is the most interesting. It is a characteristic of present knowledge that it cannot simply be negated to be gotten rid of; knowledge must be negated twice. But the tragedy of the contemporary moment is that this double negation is not, as it were, nonaligned; it is already understood as a deficiency in one's ability to imagine not utopia but dystopia: the inability to imagine that terrorists would use planes as missiles, just as it was the inability to imagine the kamikaze pilot at Pearl Harbor. These are Rumsfeld's unknown unknowns. The imagination of the future, the vision of the new, is a vision of death, fear, and terror. So not only is the unknown unknown a threat as such, and therefore difficult to bring into imagination as utopia or any another mode of thought, but the very process of attempting to imagine the unknown unknown drags into the light its opposite, the end of humanity.

Codification, Not Reification

An important future political problem will no longer be the alienation of real social relations into objects, but the extraction of abstract code *from* objects. In other words, codification not reification, is the new concern. The code in question can be either machinic or genetic; that is to say, it can be both the codification of motor phenomena or the codification of essential patterns. The vulgar Marxist approach decries the loss of authentic, qualitative social relations. But in fact the transformation under capitalism is at the same time the movement *into* the fetish of the qualitative—"Quantity has been transmuted into quality," wrote Benjamin.[33] Labor is always measured in time, in numbers. But from this numerical form, labor, comes the real, reified qualitative form of the commodity. A social relation becomes an object—this is the meaning of reification.

But when all is information, the forging of objects is no longer most important. Instead, sources, essences, recipes, and instruction sets are madly sought after and protected. The source fetishists are the new exploitative classes, what McKenzie Wark calls the "vectoralists." This is as much of a problem in genomics as it is in computer science. The practice of bioprospecting, whereby rare or unique genes are harvested from the planet's biodiversity hot spots for their value as pure information, has little by little committed entire species to digital form, ignoring and often discarding their actual lived reality.

For generations the impoverished classes have been defined as *those who have nothing but their bodies to sell.* This used to mean, simply, selling one's human labor power. Given sufficient sustenance, the impoverished classes could always manage to do this, producing at work and reproducing at home—the two requirements of workers. The dire reality of having nothing but one's body to sell has not changed. But today the impoverished classes are being exploited informatically as well as corporally. To survive, they are expected to give up not just their body's labor power but also their body's *information* in everything from biometric examinations at work, to the culling of consumer buying habits, to prospecting inside ethnic groups for disease-resistant genes. The biomass, not social relations, is today's site of exploitation.

Tactics of Nonexistence

The question of nonexistence is this: how does one develop techniques and technologies to make oneself unaccounted for? A simple laser pointer can blind a surveillance camera when the beam is aimed directly at the camera's lens. With this type of cloaking, one is not hiding, simply nonexistent to that node. The subject has full presence but is simply not there *on the screen.* It is an exploit. Elsewhere, one might go online but trick the server into recording a routine event. That's nonexistence. One's data is there, but it keeps moving, of its own accord, in its own temporary autonomous ecology. This is "disingenuous" data, or data in camouflage as not-yet-data. Tactics of abandonment are positive technologies; they are tactics of fullness. There is still struggle in abandonment, but it is not the struggle of confrontation, or the bureaucratic logic of war. It is a mode of non-existence: the full assertion of the abandonment of representation. Absence, lack, invisibility, and nonbeing have nothing to do with nonexistence. Nonexistence is nonexistence not because it is an absence, or because it is not visible, but precisely because it is full. Or rather, because it permeates. That which permeates is not arbitrary, and not totalizing, but tactical.

Of course, nonexistence has been the concern of antiphilosophy philosophers for some time. Nonexistence is also a mode of escape, an "otherwise than being." Levinas remarks that "escape is the need to get out of oneself."[34] One must always choose either being or nonbeing (or worse, becoming...). The choice tends to moralize presence, that one must be accounted for, that one must, more importantly, account for oneself, that accounting is tantamount to self-identification, to *being* a subject, to individuation. "It is this category of getting out, assimilable neither to renovation nor to creation, that we must grasp.... It is an inimitable theme that invites us to get out of being."[35] And again Levinas: "The experience that reveals to us the presence of being as such, the pure existence of being, is an experience of its powerlessness, the source of all need."[36]

Future avant-garde practices will be those of nonexistence. But still you ask: how is it possible not to exist? When existence becomes a measurable science of control, then nonexistence must become a tactic for any thing wishing to avoid control. "A being radically devoid of any representable identity," Agamben wrote, "would be absolutely irrelevant to the State."[37] Thus we should become devoid of any *representable* identity. Anything measurable might be fatal. These strategies could consist of nonexistent action (nondoing); unmeasurable or not-yet-measurable human traits; or the promotion of measurable data of negligible importance. Allowing to be measured now and again for false behaviors, thereby attracting incongruent and ineffective control responses, can't hurt. A driven exodus or a pointless desertion are equally virtuous in the quest for nonexistence. The bland, the negligible, the featureless are its only evident traits. The nonexistent is that which cannot be cast into any available data types. The nonexistent is that which cannot be parsed by any available algorithms. This is not nihilism; it is the purest form of love.

Disappearance; or, I've Seen It All Before

For Paul Virilio, disappearance is the unforeseen by-product of speed. Technology has gone beyond defining reality in the quantized frames-per-second of the cinema. Newer technologies still do that, but they also transpose and create quantized data through time stretching, morphing, detailed surface rendering, and motion capture, all with a level of resolution beyond the capacity of the human eye (a good argument for optical upgrades): "The world keeps on coming at us, to the detriment of the object, which is itself now assimilated to the sending of information."[38] Things and events are captured before they are finished, in a way, before they exist as things or events. "Like the war weapon launched at full speed at the visual target it's supposed to wipe out, the aim of cinema will be to provoke an effect of vertigo in the voyeur-traveler, the end being sought now is to give him the impression of being projected into the image."[39] Before the first missiles are launched, the battlefield is analyzed, the speeches are made, the reporters are embedded, the populations migrate (or are strategically rendered as statistical assets), and the prime-time cameras are always on. But this is not new, for many of Virilio's examples come from World War II military technologies of visualization. In this context, a person is hardly substantial—one's very physical and biological self keeps on slipping away beneath masses of files, photos, video, and a panoply of Net tracking data. But luckily you can move. All the time, if you really want to.

Hakim Bey's "temporary autonomous zone" (TAZ) is, in a way, the response to Virilio's warnings against the aesthetics of disappearance. But the issue here is nomadism, not speed. Or for Bey, nomadism is the response to speed (especially the speed produced by the war + cinema equation). A TAZ is by necessity ephemeral: gather, set up, act, disassemble, move on. Its ephemeral nature serves to frustrate the recuperative machinations of capital. The TAZ nomad is gone before the cultural and political mainstream knows what happened. This raises the issue of efficacy. The TAZ wages the risk of an efficacy that is invisible, de-presented, an efficacy whose traces are more important than the event itself. (Is this a distributed efficacy?) But this then puts us into a kind of cat-and-mouse game of forever evading, escaping, fleeing the ominous shadow of representation. Perhaps the challenge today is not that of hypervisualization (as Virilio worries), or of non-recuperation (as Bey suggests), but instead a challenge of existence without representation (or at least existence that abandons representation, a nonexistence, an a-existence). "Disappearance is not necessarily a 'catastrophe'—except in the mathematical sense of 'a sudden topological change.'"[40] And so goes the juvenile interjection of apathy, only now reimagined as distinctly tactical and clever: whatever.

Stop Motion

First call to mind the stories of H. P. Lovecraft, or perhaps Elias Merhinge's film *Begotten*. A person comes across a lump of gray, dirty clay. Just sitting there. No, it is starting to move, all by itself. It makes squishy sounds as it does so. When it's finished it has formed itself into the face of the person, and the person is suddenly Dr. Faustus. Or take another scenario: a person comes across a strange, part-insect, part-amphibian thing lying there. Is it alive? How can one be sure? Poke it, carefully nudge it, maybe even touch it. Or the grainy, dirty body lying in the mud can't stop convulsing, and yet it is dead. The traditions of supernatural horror and "weird fiction" are replete with scenarios like these, populated by "unnameable horrors," a "thing on the doorstep," unidentified "whisperers in darkness," and a "ceaseless, half-mental calling from the underground."[41]

The question of animation and the question of "life" are often the same question. Aristotle's *De anima* identified motion and animation as one of the principal features of living beings: "Now since being alive is spoken of in many ways ... we may say that the thing is alive, if, for instance, there is intellect or perception or spatial movement and rest or indeed movement connected with nourishment and growth and decay."[42] If it moves, it is alive. But the mere fact of movement isn't enough. The Aristotelian notion of substance implies that there must be some principle of self-movement beyond the mere matter of the thing, a particular form or principle of actualization. Aquinas would take this in a more theological vein as implying the necessity of a first mover, and movement that makes all subsequent movement possible. "It should be said that it is necessary that all agents act for the sake of an end.... But the first of all causes is the final cause.... However, it should be noted that a thing tends to the end by its own action or motion in two ways. In one way, as moving itself to the end, as man does; in another way, as moved to the end by another, as the arrow tends to a definite end because it is moved by the archer."[43] Strangely enough, in a theological treatise, we now have technology as part of the equation (and an argument for or against artificial life).

But this type of thinking presupposes a certain relation between thing and movement, object and trajectory. It assumes that first, the object exists, and second, that it is then moved. In technological terms, we would say that first there are "nodes" and then there are "edges" or links between nodes. In political terms, we would say that first there is a subject, and then the subject has agency.

What about the reverse? "There are changes, but there are underneath the change no things which change: change has no need of a support."[44] This reversal makes a difference ontologically, but the question is whether it makes a difference technologically and politically. This is the conundrum of contemporary debates over network forms of organization, be they netwars, wireless communities, or file sharing. Graph theory—the very mathematical roots of network science—begins from the classical division between node and edge *and in doing so privileges space over time, site over duration.* Nevertheless, it is hard to see how networks can be thought of—technically or politically—in any other way. Except, perhaps, if movement or animation becomes ontological, as it is in Heraclitus, or even if it is considered a universal right.

Consider Raoul Vaneigem's proposal: "No one should be obstructed in their freedom of movement.... The right to nomadism is not a passive migration, decided by poverty or scarcity among producers and consumers, but the wandering of individuals aware of their creativity and concerned not to fall into a condition of dependency and be objects of charity."[45] Vaneigem's "wandering body" reserves for human beings even "the right to stray, to get lost and to find themselves."

Perhaps in the ontological inversion of graph theory and networks a new kind of "politics of desertion" will be discovered.

Pure Metal

A theme often repeated in poststructuralist thought is the so-called decentering of the liberal-humanist subject, a trend that led to charges of antihumanism against such thinkers. There is no presocial, universal, or essential self, one is told; there is only the play of surface effects and signs, themselves largely determined by social, economic, and political forces. The question is not "who am I?" but rather "how did an 'I' become so central to our social, economic, and political existence?" This cultural constructionism therefore focuses less on an innate self or identity than on the ways that gender/sexuality, race/ethnicity, and class/status form, shape, and construct a self.

In this debate over the decentering or the recentering of the human subject, we wonder: what of the *nonhuman*, that which seemingly does not concern us at all? The term "nonhuman" has in fact been used in science studies and media studies, for example, by Bruno Latour. Focusing on knowledge production in the sciences, Latour notes that any scientific experiment involves not only human investigators but a range of "nonhuman actants" (rather than human actors) that may include laboratory technologies, organisms, materials, research grants, institutions and corporations, and so on. Latour gives a simpler, if more mundane, example: the speed bump. Speed bumps do nothing on their own; they simply exist as inert matter. Yet their presence affects our human actions (slowing down, swerving). Human made, yes, but their being crafted by humans (or, more appropriately, machines) says nothing about their interactions with human beings and their integration into the everydayness of the human world. This integration leads Latour to speak of a "parliament of things," a "nonmodern Constitution" in which even "things" seem to be democratic. Yet despite Latour's emphasis on the way that human actions are influenced by nonhuman actants, there is a sense in which the nonhuman is still anthropomorphized, a sense in which the circumference of the human is simply expanded.

If Latour gives us a paradoxically anthropomorphic version of the nonhuman, then Jean-François Lyotard gives us the opposite: the absolute limit of the human, a quantum, even cosmic, nonhuman. Lyotard prefers the term "inhuman," but we detect the same debate in his later writings. Lyotard suggests that one does not need to theorize the nonhuman, for a process of "dehumanization" has already been taking place for some time via postmodern technology and postindustrial capitalism. An inhumanity is produced in which all politics is reduced to voting, all culture to fashion, all subjectivity to the stratifications of class, and so on. The only response, he argues, is to discover another kind of inhuman, one that is a "no-man's land," one that explores another type of transformation, a transformative transformation (rather than a change that produces the same). "Humanity is only human if people have this 'no-man's land.'"[46] By definition, this potential site of resistance cannot be named unless it is discovered. For Lyotard, this is the domain of our contemporary "sublime" in the arts, the very site of discovering that which is beyond representation, that which lies beyond human thought itself.

Between these two approaches to the nonhuman, an incorporating and a discorporating one, the nonhuman continues to be negatively defined: the human "minus" something, or "what the human is not." One always begins from the human and then moves outward. But what of a nonhuman within the human, just as the swarm may emanate from within the network? Or better, what of a nonhuman that *traverses* the human, that *runs through* the human? Would this not be the "matter" of the nonhuman? By contrast, the matter of the human is hylomorphic, a set of matters that have their end in a given form, matter-into-form, form as the *telos* of matter. This is the human view of technology, instrumentality as the ability literally to shape the world. The inverse of this would be a metallomorphic model, of which metallurgy is the privileged example. "Matter and form have never seemed more rigid than in metallurgy; yet the succession of forms tends to be replaced by the form of a continuous development, and the variability of matters tends to be replaced by the matter of a continuous variation."[47] Melting, forging, quenching, molding—metallurgy operates according to different relations of matter-form. Furthermore, metallurgy, or the idea of *un métal pur*, has nothing to do with the forging of swords or coins per se. "Not everything is metal, but metal is everywhere." There is a metallurgy that cuts across all complex matter, a sort of contagion of metallomorphic patterns, recombinations of bits and atoms, a metalmorphosis of living forms that is not less ambivalent for its being metallurgic: "The huge population of viruses, combined with their rapid rates of replication and mutation, makes them the world's leading source of genetic innovation: they constantly 'invent' new genes. And unique genes of viral origin may travel, finding their way into other organisms and contributing to evolutionary change."[48]

The Hypertrophy of Matter (Four Definitions and One Axiom)

Definition 1. *Immanence* describes the process of exorbitance, of desertion, of spreading out. Networks cannot be thought without thinking about immanence (but not all networks are distributed networks). Immanence is not opposed to transcendence but is that which distributes in transcendence. Immanence is formally not different from self-organization. Immanence is ontologically not different from what Spinoza describes as *causa sui*.

Definition 2. *Emptiness* is an interval. Emptiness is "the space between things," or what in graph theory is called an edge. Emptiness is the absence of space, as if in itself. Emptiness is always "$n-1$." It is the pause that constitutes the network.

Definition 3. *Substance* is the continual by-product of the immanence of emptiness. Substance is the effect, not the cause, of networks, akin to what graph theory calls a node. Substance is the point at which monism and pluralism implode. Substance is the indistinction of the one and the many, the production of the nodes that constitute a network.

Definition 4. *Indistinction* is the quality of relations in a network. Indistinction is the "third attribute" never postulated by Spinoza (thought, extension, indistinction). *Indistinction* is not *nondistinc*-

tion. Nondistinction effaces distinctions, whereas indistinction proliferates them. Indistinction is the ability to autogenerate distinctions recursively.

Axiom 1. Networks have, as their central problematic, the fact that they prioritize nodes at the same time as they exist through the precession of edges. Networks are less mystical "first causes," and more the production of topological conditions and possibilities.

The User and the Programmer

Freedom of expression is no longer relevant; freedom of use has taken its place. Consider two categories: the computer user and the computer programmer. One designates the mass of computer society, the other a clan of technical specialists. Or not? The user and the programmer are also two rubrics for understanding one's relationship to art. ("There are two musics," wrote Roland Barthes, "the music one listens to, [and] the music one plays.")[49] "User" is a modern synonym for "consumer." It designates all those who participate in the algorithmic unfoldings of code. On the other hand, "programmer" is a modern synonym for "producer." It designates all those who participate *both* in the authoring of code and in its process of unfolding. Users are executed. But programmers execute themselves. Thus "user" is a term for any passive or "directed" experience with technology, while "programmer" means any active or "undirected" experience with technology. Taken in this sense, anyone can be a programmer if he or she so chooses. If a person installs a game console modchip, he is programming his console. If she grows her own food, she is programming her biological intake.

The unfortunate fallout of this is that most legal prohibitions are today migrating away from prohibitions on being (the user model) toward prohibitions on doing (programmer). Today there are more and more threats to programming in everyday life: digital rights management agreements prohibit specific uses of one's purchased property; sampling has become a criminal act. Hence future politics will turn on freedoms of use, not on the antiquated and gutted freedom of expression.

Fork Bomb IV

```
#!/usr/bin/perl
while (rand()<.9) {
  if(fork && push(@O,$_)) {
    print "_"x@O, crypt($O[$#O],$#O);
  } else {
    print " | "x@O;
  }
}
```

Interface

Define "interface" like this: an artificial structure of differentiation between two media. What is a structure of differentiation, and why artificial? Differentiation happens whenever a structure is added to raw data, when mathematical values are parted and inflected with shape. These eight binary digits are different from those eight binary digits; or this word is a markup tag and that word is plain text; or this glob is a file and that glob is an executable. It is artificial in the sense that it is made. That is to say, data does not come into being fully formed and whole but instead is inflected with shape as the result of specific social and technical processes. Interface is how dissimilar data forms interoperate. In fact, if two pieces of data share an interface, they are designed *only* to interoperate. Classes in object-oriented programming have interfaces. Hypertext transfer protocol (http) is the interface between Web server and client. At the same time, http shares an interface with TCP (they are different groups of bytes, yet the latter interfaces with the former). There is a grammar of articulation between virus and host that is an interface. Interface is the difference sculpted from what George Boole called the "Nothing" (zero) and the "Universe" (one).

There Is No Content

Theories of media and culture continue to propagate an idea of something called "content." But the notion that content may be separated from the technological vehicles of representation and conveyance that supposedly facilitate it is misguided. Data has no technique for creating meaning, only techniques for interfacing and parsing. To the extent that meaning exists in digital media, it only ever exists as the threshold of mixtures between two or more technologies. Meaning is a data conversion. What is called "Web content" is, in actual reality, the point where standard character sets rub up against the hypertext transfer protocol. There is no content; there is only data and other data. In Lisp there are only lists; the lists contain atoms, which themselves are other lists. To claim otherwise is a strange sort of cultural nostalgia, a religion. Content, then, is to be understood as a relationship that exists between specific technologies. Content, if it exists, happens when this relationship is solidified, made predictable, institutionalized, and mobilized.

Trash, Junk, Spam

Trash, in the most general sense, implies the remnants of something used but later discarded. The trash always contains traces and signatures of use: discarded monthly bills, receipts, personal papers, cellophane wrapping, price tags, spoiled food. And so much more: trash is the set of all things that have been cast out of previous sets. It is the most heterogeneous of categories, "all that which is not or no longer in use or of use."

Junk is the set of all things that are not of use at the moment, but may be of use someday, and certainly may have been useful in the past. Junk sits around, gathering dust, perhaps moved occasionally from one location to another. It may be of some use someday, but this use is forever unidentified. Then in an instant it is no longer junk but becomes a spare part for a car, a new clothing fashion, or an archive of old magazines. "Don't throw that out, it might come in handy someday."

Is spam trash or junk? Spam e-mails are thrown away, making them trash. And there is so-called junk e-mail, a name borrowed from junk mail, both of which are typically cast off as trash. But spam is something entirely different.

Spam is an exploit, and an incredibly successful one. One spends hours crafting the perfect algorithmic e-mail filter to catch the offending spam while still permitting all "meaningful" e-mail to penetrate safely. And still it is a foregone conclusion that each e-mail session will require a certain amount of manual spam identification and deletion. Spam is not quite as aggressive as a computer virus or an Internet worm, and though spam attachments can be viruses or worms, spam is by and large something to be deleted or marked as trash. As an informational entity, spam is less a question of antivirus protection and more one of bureaucratic data management: algorithmic filtering of the meaningful from the meaningless, marking the corralled messages for deletion, junking attachments from mailboxes, and approving or denying the status of one message over another. Spam leverages the low marginal costs of electronic mail by exploiting the flaws (some would say features) of the planetary e-mail network and in doing so elicits an antagonistic response from users in the form of informatic network management.

Spam signifies nothing *and yet is pure signification.* Even offline junk mail, anonymously addressed to "Current Resident," still contains nominally coherent information, advertising a clearance sale or fast-food delivery. Spam is not anonymous, for a receiver address is a technical requirement, and yet its content has no content, the receiver address the result of algorithmic data collection and processing done by Web spiders and collected in massive databases. Often spam uses Web bugs that call back to a central server, confirming the existence of the receiver's address, nothing more, nothing less. The spam might be deleted, but the damage is done. A subject line might advertise one thing—typically the three P's, porn, pharmaceuticals, and payment notices—but often the body of the e-mail advertises something else entirely. Misspellings and grammatical errors are strategic in spam e-mails, in part to elude spam filters in an ever-escalating game of syntactic hide-and-seek. Thus Ambien becomes "Amb/en," or Xanax becomes "X&nax." Many spam generators use keywords from e-mail subject headings and recombine those terms into new subject headings. But in the end, spam e-mail simply wants to generate a new edge for the graph; it wants the user to click on a URL, or to open an attachment, either action a new link in the net.

In the midst of all this, something has happened that may or may not have been intentional. Spam e-mails, with their generated misspellings, grammatical errors, and appropriated keywords and names, have actually become generative in their mode of signification. But this generativity has nothing to do with any direct relation between signifier and signified. It is what Georges Bataille called a "general economy" of waste, excess, and expenditure, except that this excess is in fact produced and managed informatically by software bots and e-mail filters.[50] Linguistic nonsense is often the result, a grammatical play of subject headings that would make even a dadaist envious: "its of course grenade Bear" or "It's such a part of me I assume Everyone can see it" or "Learn how to get this freedom . . ." Spam-bots are the heirs of the poetry of Tristan Tzara or Hugo Ball. Spam is an excess of signification, a signification without sense, precisely the noise that signifies nothing—except its own networked generativity.

Coda: Bits and Atoms

Networks are always exceptional, in the sense that they are always related, however ambiguously, to sovereignty.

This ambiguity informs contemporary discussions of networks and the multitude, though in a different fashion. Hardt and Negri, for instance, describe the multitude as a "multiplicity of singularities," a group that is unified by "the common" but remains heterogeneous in its composition. The multitude is, in their formulation, neither the centralized homogeneity described by Hobbes nor the opposite condition of a purely digressionary chaos. "The concept of the multitude rests on the fact, however, that our political alternatives are not limited to a choice between central leadership and anarchy."[1]

Hardt and Negri find indications of such a multitude in the worldwide demonstrations against the WTO, in the different Latin American peasant revolts, and in the tradition of the Italian "workerism" movement. For them, such examples offer a hint of a type of political organization that resists the poles of either centrist sovereignty or centerless anarchy:

> To understand the concept of the multitude in its most general and
> abstract form, let us contrast it first with that of the people. The
> people is one.... The people synthesizes or reduces these social dif-
> ferences into one identity. The multitude, by contrast, is not unified
> but remains plural and multiple. This is why, according to the domi-
> nant tradition of political philosophy, the people can rule as a
> sovereign power and the multitude cannot. The multitude is com-
> posed of a set of singularities—and by singularity here we mean a
> social subject whose difference cannot be reduced to sameness, a
> difference that remains different.... The plural singularities of the
> multitude thus stand in contrast to the undifferentiated unity of
> the people.[2]

These terms—the one and the many, sovereignty and multitude—are
at once ultracontemporary and at the same time resolutely historical.
Paolo Virno, for instance, notes that such debates have evoked a "new
seventeenth century," in which the face-off between Hobbes and Spinoza comes into the foreground.[3] Virno nuances the opposition between
the one and the many and suggests that the contemporary multitude
is opposed to the very opposition itself of the one and the many:

> And it is precisely because of the dissolution of the coupling of these
> terms, for so long held to be obvious, that one can no longer speak
> of a *people* converging into the unity of the state. While one does
> not wish to sing out-of-tune melodies in the post-modern style
> ("multiplicity is good, unity is the disaster to beware of"), it is
> necessary, however, to recognize that the multitude does not clash
> with the One; rather, it redefines it. Even the many need a form of
> unity, of being a One. But here is the point: this unity is no longer
> the State; rather, it is language, intellect, the communal faculties of
> the human race. The One is no longer a *promise,* it is a *premise.*[4]

The multitude has a focus, a direction, but its actions and decisions
are highly distributed. The "One" of the multitude is less a transcendent "One," serving to homogenize a collectivity, and more like an
immanent "One" (we would do better to say a "univocity") that is
the very possibility of collective organization.

Gone are the days of centralized, uniform mass protests; instead
one witnesses highly distributed, tactical modes of dissent that often
use high and low forms of technology. The very fact that the multitude is not "One" is its greatest strength; the multitude's inherently
decentralized and even distributed character gives it a flexibility and

robustness that centralized modes of organization lack. In fact, the analyses of Hardt and Negri, as well as Virno, seem to imply the identity between the formal and ideological planes of the multitude, an isomorphism between the topological and political levels.

Contemporary analyses of "multitude" (such as those of Hardt and Negri, and Virno) share significant affinities with Arquilla and Ronfeldt's analysis of "netwar." In this intersection, political allegiances of Left and Right tend to blur into a strange, shared concern over the ability to control, produce, and regulate networks.

For Arquilla and Ronfeldt, "netwar refers to an emerging mode of conflict (and crime) at societal levels, short of traditional military warfare, in which the protagonists use network forms of organization and related doctrines, strategies, and technologies attuned to the information age."[5] While they write from a perspective far to the right of Hardt and Negri or Virno, Arquilla and Ronfeldt do acknowledge that netwar has two sides to it: "We had in mind actors as diverse as transnational terrorists, criminals, and even radical activists."[6] While they, like Hardt and Negri, discuss the 1999 anti-WTO demonstrations in Seattle, they also discuss international terrorist networks, the Zapatista movement, and pro-democracy uses of the Internet in Singapore, Vietnam, and Burma. Thus "netwar can be waged by 'good' as well as 'bad' actors, and through peaceful as well as violent measures."[7] For Arquilla and Ronfeldt, the diversity of types of netwar, as well as their complexity (they outline five distinct levels at which netwar operates: technological, doctrinal, ideological, narratological, and social), makes them a unique emerging form of political action, with all the unknowns that such forms imply. For them, "netwar is an ambivalent mode of conflict."[8] And as we discussed at the outset, the radicality of Arquilla and Ronfeldt's analyses is to suggest that "it takes networks to fight networks"—that network modes of action require network-based responses (a suggestion that, to be sure, works against many of the U.S. military's legacy centrist hierarchies).

Despite their political differences, both the concept of the multitude and the concept of netwars share a common methodological approach: that "the 'unit of analysis' is not so much the individual as it is the network in which the individual is embedded."[9]

When Hardt and Negri talk about the multitude as an emerging "social subject," they imply a view of the whole that does not reduce it to a uniform, homogeneous unit, just as Arquilla and Ronfeldt's analyses understand netwars as connected by a common vision but deployed according to diverse and divergent means.

Moreover, Hardt and Negri are deliberately ambiguous when it comes to the political status of distributed networks: they use the distributive diagram to explain empire and the multitude alike.[10] So while form is important (e.g., is a network distributed or centralized?), it is not the most important factor for evaluating different movements. In Hardt and Negri one must examine the content of any distributed network to determine its political effects. "We have to look not only at the form but also the content of what [movements] do," they remind us.

Hardt and Negri's argument is never that distributed networks are inherently resistive. The network form is *not* tied to any necessary political position, either progressive or reactionary.[11] In fact, this is the primary reason why there can exist network-to-network struggle. Both the forces of the multitude and the counterforces of empire organize themselves around the topology of the distributed network; there is no bonus given to either side simply for historically adopting the distributed form.[12] So at the finish of *Empire* and *Multitude*, we end up in a symmetrical relationship of struggle. Empire is a distributed network, and so is the multitude. In fact, the very notion that networks might be in a relationship of political opposition at all means that networks *must* be politically ambiguous.

But what we suggest is missing from both books is any vision of a *new future of asymmetry*. Slow, deliberate reform comes about through the head-to-head struggle of like forces. But revolutionary change comes about through the insinuation of an asymmetrical threat.[13] So the point is that if today one agrees that a new plateau of global symmetry of struggle exists in the world—networks fighting networks, empire struggling against the multitude—then what will be the *shape* of the new revolutionary threat? What will be the undoing of the distributed network form, just as it was the undoing of a previous one? From where will appear the anti-Web? And what will it look like? Resistance *is* asymmetry—and this is where we part ways with Hardt and Negri—formal sameness may bring about reform, but formal incommensurability breeds revolution.

Because both empire and the multitude employ the distributed network form, it is not sufficient to remain politically ambiguous on the question of distributed networks. A decision has to be made: we're tired of rhizomes. One must not only analyze how distributed networks afford certain advantages to certain movements; one must critique the logics of distributed networks themselves. Many political thinkers today seem to think that "networked power" means simply the aggregation of powerful concerns into a networked shape, that networked power is nothing more than a network of powerful individuals. Our claim is entirely the opposite, that the materiality of networks—and above all the "open" or "free" networks—exhibits power relations regardless of powerful individuals.

At this point, we pause and pose a question: Is the multitude always "human"? Can the multitude or netwars not be human and yet still be "political"? That is, are individuated human subjects always the basic unit of composition in the multitude? If not, then we must admit that forms such as the multitude, netwars, and networks exhibit unhuman as well as human characteristics.

We mean such questions less as an issue about agency (the issue of how to instrumentalize networks, how to use them as tools), and more as an issue about the nature of constituent power in the age of networks (the issue of the kinds of challenges that networks pose to the way we think about "politics"). If, as the saying goes, networks operate at a global level from sets of local interactions, and if one of the defining characteristics of networks is the way in which they redefine "control" in the Deleuzian sense, then we are moved to ask what *inhabits* the gap left open by the limitations of autonomous, causal, human agency. Nothing about networks leads us to believe that they are inherently egalitarian forms—a tendency displayed in the numerous popular-science books on networks. Not all networks are equal, to be sure. Furthermore, networks often display asymmetrical power relationships (as in the "directed" graph). But if no *one* controls the network in an instrumental sense, if there is indeed a defacement of enmity, then how do we account for and "live against" such differences and asymmetries?

Our suggestion may at first seem perplexing. We suggest that the discussions over the multitude, netwars, and networks are really discussions about the unhuman within the human.

By using the word "unhuman," we do not mean that which is against the human or is antihuman. Nor do we mean the use of cybernetic technology to evolve beyond the human. This is not a nihilism (antihumanism) or a technophilia (posthumanism). We do not deny the crucial role that human action and decision play, even at the most micropolitical, localized level. But we do wonder if the thinking about these so-called emerging forms really goes far enough in comprehending them.

Difficult, even frustrating, questions appear at this point. If no single human entity controls the network in any total way, then can we assume that a network is not controlled by humans in any total way? If humans are only a part of a network, then how can we assume that the ultimate aim of the network is a set of human-centered goals?

Consider the examples of computer viruses or Internet worms, of emerging infectious diseases, of marketing strategies employing viral marketing or adware, of the unforeseen interpersonal connections in any social network, of the connections between patterns of immigration and labor in the United States, of the scaling up of surveillance in U.S. Homeland Security and the Patriot Act, of the geopolitics of the Kyoto Treaty and climate change. At the macro and micro levels, it is not difficult to note at least some element in every network that frustrates total control—or even total knowledge.

In fact, it is the very idea of "the total" that is both promised and yet continually deferred in the "unhumanity" of networks, netwars, and even the multitude.

The point here is not that networks are inherently revolutionary but that networks are constituted by this tension between unitary aggregation and anonymous distribution, between the intentionality and agency of individuals and groups on the one hand, and the uncanny, unhuman intentionality of the network as an "abstract" whole.

The network is this combination of spreading out and overseeing, evasion and regulation. It is the accident and the plan. In this sense, we see no difference between the network that works too well and the network that always contains exploits.

Of course, from another perspective, there is a great difference between the network that functions well and the network that fails— from "our" point of view. And this is precisely why the examples discussed earlier, such as Internet worms or emerging infectious diseases, evoke a great deal of fear and frustration.

Perhaps we have not paid enough attention to the "elemental" aspects of networks, netwars, or the multitude. Perhaps the most interesting aspects of networks, netwars, and the multitude are their unhuman qualities—unhuman qualities that nevertheless do not exclude the role of human decision and commonality. This is why we have always referred to protocol as a physics. Networks, generally speaking, show us the unhuman in the human, that the individuated human subject is not the basic unit of constitution but a myriad of information, affects, and matters.

For this reason, we propose something that is, at first, counterintuitive: to bring our understanding of networks to the level of bits and atoms, to the level of aggregate forms of organization that are material and unhuman, to a level that shows us the unhuman in the human.

What exactly would such an unhuman view of networks entail?[14] We close—or rather, we hope, open—with a thought concerning networks as "elemental" forms. By describing networks as elemental, we do not mean that our understanding of networks can wholly be reduced to physics, or a totally quantitative analysis of bits and atoms. Nevertheless we find in the bits and atoms something interesting, a level of interaction that is both "macro" and "micro" at once.

The level of bits and atoms suggests to us not modern physics or postmodern computing but something totally ancient—an ancient, even pre-Socratic understanding of networks. The pre-Socratic question is a question about the fabric of the world. Of what is it made? What is it that stitches the world together, that links part to part in a larger whole? The answers given, from Thales to Anaxagoras, involve the elemental. Water, fire, air, "mind," or some more abstract substance... Heraclitus, for instance, gives us a world in which everything flows—empires rise and fall, a person remains a person throughout youth and old age, and one can never step into the same river twice. For Heraclitus, it is fire that constitutes the world. But he does not mean "fire" as a denotated thing, for the flame or the sun itself point to another "fire," that of dynamic morphology, a propensity of energentic flux. This kind of fire is more elemental than natural. The same can be said for Parmenides, who is now more commonly regarded as the complement, rather than the opposite, of Heraclitus. The emphasis on the "One"—the sphere without circumference—leads Parmenides to the fullness of space, a plenum that emphasizes the interstitial aspects of the world. If everything flows (the statement of Heraclitus), then all is "One" (the proposition of Parmenides).

A movement between a world that is always changing and a world that is immobile, between a world that is always becoming and a world that is full—the movement and the secret identity between these positions seem to describe to us something fundamental about networks. Networks operate through ceaseless connections and disconnections, but at the same time, they continually posit a topology. They are forever incomplete but always take on a shape.

The shape also always has a scale. In the case of certain network topologies such as the decentralized network, the scale is fractal in nature, meaning that it is locally similar at all resolutions, both macroscopic and microscopic. Networks are a matter of scaling, but a scaling for which both the "nothing" of the network and the "universe" of the network are impossible to depict. One is never simply inside or outside a network; one is never simply "at the level of" a network. But something is amiss, for with fields such as network science and new forms of data visualization, attempts are made to image and manage networks in an exhaustive sense. The impossibility of depiction is ignored, and the network is *imagined* nonetheless. Accidents, failures, and exploits, both imaginative and material, are part and parcel of any network. These are strange and often bewildering kinds of accidents and failures—the accidents that are prescribed by the design, the failures that indicate perfect operation.

Networks are elemental, in the sense that their dynamics operate at levels "above" and "below" that of the human subject. The elemental is this ambient aspect of networks, this environmental aspect— all the things that we as individuated human subjects or groups do not directly control or manipulate. The elemental is not "the natural," however (a concept that we do not understand). The elemental concerns the variables and variability of scaling, from the micro level to the macro, the ways in which a network phenomenon can suddenly contract, with the most local action becoming a global pattern, and vice versa. The elemental requires us to elaborate an entire climatology of thought.

The unhuman aspects of networks challenge us to think in an elemental fashion. The elemental is, in this sense, the most basic and the most complex expression of a network.

As we've suggested in this book, networks involve a shift in scale, one in which the central concern is no longer the action of individuated agents or nodes in the network. Instead what matters more and more is the very distribution and dispersal of action throughout the network, a dispersal that would ask us to define networks less in terms of the nodes and more in terms of the edges—or even in terms other than the entire, overly spatialized dichotomy of nodes and edges altogether.

In a sense, therefore, our understanding of networks is *all-too-human* . . .

Appendix

Notes for a Liberated
Computer Language

"To record the sound sequences of speech," wrote Friedrich Kittler, "literature has to arrest them in a system of twenty-six letters, *thereby categorically excluding all noise sequences*."[1] A fascinating act of transduction is language. But we worry. We worry about the imaginary, supplemental alphabets starting with letter twenty-seven. This is the impulse behind our notes for a liberated computer language, to re-introduce new noisy alphabets into the rigid semantic zone of informatic networks. The symbolic economies discussed in the 1970s by theorists such as Jean-Joseph Goux have today been digitized and in-stantiated into the real codes of life itself. What was once an abstract threat, embodied in specific places (the school, the factory) with particular practices of control and exploitation, is today written out in gross detail (the RFCs, the genome), incorporated into the very definitions of life and action. This is why liberated languages are so important today. We consider there to be little difference between living informatic networks and the universal informatic languages and standards used to define and sculpt them. If the languages are finite, then so, unfortunately, are the life possibilities. Thus a new type of language is needed, a liberated computer language for the articulation of political desires in today's hostile climate of universal informatics. We offer these notes for a liberated computer language as a response to the new universalism of the informatic sciences that have subsumed all of Goux's symbolic economics.

Most computer languages are created and developed according to the principles of efficiency, utility, and usability. These being but a fraction of the human condition, the following language specification shuns typical machinic mandates in favor of an ethos of creative destruction. The language contains data types, operators, control structures, and functions, the latter defined using a standard verb-object syntax adopted from computer science whereby the function name appears first followed by the variable being passed to the function (example: functionName *VARIABLE*).

Data Types

creature   an entity that is not readable, writable, or executable but that exists

doubt   an entity that questions its own status as an entity, which it may or may not actually be (see also Denial and Refusal)

empty   a null entity lacking any and all material or immaterial distinction. The empty type allows for dynamic creation of new, unimagined types at runtime.

flaw   a fault or imperfection associated with another entity

flip   an entity that oscillates between two other data types

full   contains the complete universe of all information and matter

gateway   an associative entity connecting two or more other entities

glossolalia   an entity that is readable, writable, or executable only on a hypothetical machine

incontinent   an entity that involuntarily expresses itself as any other data type

infinity   an entity that is unbounded in quality and magnitude

palimpsest   an entity that contains the traces of a previous entity

poltergeist   an entity whose sole function is to invoke another
                 entity

putrefaction   an entity that only produces its data when deleted

qualitative   an entity with contents that are not numerical in
                 value

random   expresses a random entity from a random type

topology   an arrangement of interrelated, constituent parts

unknown   an entity that cannot be specified, identified, or
                 evaluated in any intelligible way

vector   a compound type representing intensity and
                 direction, consisting of an origin type and a
                 destination type

whatever   an entity that always matters. It is not defined as part
                 of any set (including the set of "whatevers") and
                 cannot be identified through reference to either the
                 particular or the general.

zombie   a process that is inactive but cannot be killed

Operators

( )   cast   transform an entity from its current type to a
                      new type

– –   debase   decreases the political or social standing of
                      the predicate

!=   disassignment   assigns any other entity except for the one
                      specified in the predicate

=/   disputatio   spontaneously makes further, perhaps useless,
                      distinctions within a given entity

::   figuration   establishes a figurative relationship between
                      two or more entities

–	hybrid	combines two or more entities into a new hybrid entity cast from the "empty" type
?+	manna	assigns new values to preexisting entities, the meaning of which is known only to the machine
<>	negotiate	reassigns the predicate entity based on a negotiation between it and the subject entity
" "	normative	attaches a political evaluation to a type or code block
∧	parasite	establishes a parasitical relationship between two or more entities
++	privilege	increases the political or social standing of the predicate

Control Structures

boredom	executes a code block in a trivial, meaningless way that often results in dull, tedious, or sometimes unexpected output
exceptional	designates an abnormal flow of program execution and guarantees that it will never be handled as an error
flee	a branching construct that moves flow control from the current instruction to a stray position in the program
historic	executes a code block by evaluating an entity according to its current value as well as all previous values
maybe	allows for possible, but not guaranteed, execution of code blocks
never	guarantees that a block of code will never be executed. This is similar to block quotes in other languages, except that "never" blocks are not removed during compilation.
potential	evaluates an entity only according to as-yet-unrealized possibility

singular   evaluates an entity in a manner that does not consider the entity's membership in any set or as a representative of any universal quality

unordered   executes a set of statements out of sequence

vitalize   endows any entity with ineffable, unpredictable, unexplained characteristics

Functions

backdoor ENTITY   installs a backdoor in the specified entity. If no target is provided, the backdoor is installed in the local entity.

bandwidth AMOUNT   enlarges or reduces communication bandwidth by AMOUNT

bitflip DATA, NUMBER   randomly flips a specified number of bits in a digital source specified by DATA

bug PROGRAM, NUMBER   introduces specified NUMBER of bugs into the code of the specified program

crash TIME   crashes the machine after the number of seconds provided by TIME. If TIME is not provided, the crash will occur immediately.

degrade HARDWARE, TIME   introduces wear and tear, specified by number of months given in TIME, into specified HARDWARE

desert HOST   a sudden, apparently random cessation of all functions, tasks, and processes. The departure of the operating system from the machine.

destroy ENTITY   eliminates the specified entity

disidentify *ENTITY*	removes all unique IDs, profile data, and other quantitative identifiers for the specified entity
drift *ENTITY*	initiates for specified *ENTITY* an aimless wandering within or between hosts
emp *TIME*	after the number of seconds provided by *TIME*, this function sends an electromagnetic pulse, neutralizing self and all machines within range
envision	an inductive function for articulation of unknown future realities. Often used in conjunction with rebuild.
exorcise *USER*	prohibits *USER* from accessing a given host. If *USER* is not specified, function stipulates that the host will run only when there are no users present.
fail *FUNCTION*	introduces logical fallacies into any other language method specified by *FUNCTION*
frees *TIME*	frees the machine from operating by freezing it for the number of seconds specified in *TIME*
invert *HEX*	allows a machine to infect itself with malicious code specified in *HEX* by first sending that code across a network to other machines and then receiving it back in an altered form
jam *NETWORK*	sends jamming signal to the specified *NETWORK*

lose *DEVICE*	unlinks a random file on the storage medium specified by *DEVICE*
mutate *SEQUENCE*	introduces a mutation into the given informatic *SEQUENCE*
narcolepsis *HOST*	unexpectedly initiates "sleep" mode in a given organic or inorganic *HOST*
netbust *TARGET*	exposes a network specified in *TARGET* to extremely high voltages, thereby fatally damaging any network hardware attached to the network. *TARGET* can also be "self" to affect only the local interface.[2]
noise *PROTOCOL, AMOUNT*	scatters a specific *AMOUNT* of random noise packets into the default network interface using the specified *PROTOCOL*
obfuscate *SEQUENCE*	renders any given *SEQUENCE* (gene, character string, etc.) illegible to any known parsing technologies
obsolesce *HARDWARE*	renders any given piece of *HARDWARE* obsolete. Opposite of reclaim.
overclock *MULTIPLIER*	increases the clock frequency of the central processing unit according to the value of *MULTIPLIER*. A negative value will decrease the clock frequency.
possess *SEQUENCE*	allows any system, living or nonliving, to take control of itself
processKill	selects a process at random and kills it

processScramble	randomly renumbers all currently running process IDs
rebuild *ENTITY*	begins the process of rebuilding the specified entity. Often used to remedy the effects of destroy.
reclaim *HARDWARE*	rescues any given piece of *HARDWARE* from obsolescence. Opposite of obsolesce.
reject	rebuffs the current state of affairs. Often used as a precursor to destroy.
repress *MEMORY*	disables any and all attempts to anthropomorphize the machine
reverseEngineer *TARGET*	If object specified in *TARGET* is an application, this function decompiles the application and returns commented source code. If the object specified in *TARGET* is a protocol, this function returns a formal description of the protocol.
scramble *DEVICE*	randomly shuffles all filenames on the storage medium specified by *DEVICE*
selfDestruct	imposes fatal physical damage on self. Equivalent to destroy *SELF*.
struggle	assists agitation and opposition to existing exploitation and control
zapMemory	clears all RAM on local machine

Notes

Prolegomenon

1. For more on the dialogue, see Geert Lovink and Florian Schneider, "Notes on the State of Networking," *Nettime*, February 29, 2004; and our reply titled "The Limits of Networking," *Nettime*, March 24, 2004.

2. This is seen in books like Bob Woodward's *Plan of Attack*.

3. Pit Schultz, "The Idea of Nettime," *Nettime*, June 20, 2006.

4. Giorgio Agamben, *The Coming Community* (Minneapolis: University of Minnesota Press, 1993), 85.

5. It's important to point out that terms such as "postmodernity" or "late modernity" are characterized less by their having broken with or somehow postdated modernity, but instead exist in a somewhat auxiliary rapport with modernity, a rapport that was never quite a break to begin with and may signal coincidence rather than disagreement. Fredric Jameson's book *A Singular Modernity* (London: Verso, 2002) plots this somewhat confusing boomerang effect.

6. Michael Hardt and Antonio Negri, *Multitude: War and Democracy in the Age of Empire* (New York: Penguin, 2004), 62.

7. John Arquilla and David Ronfeldt, "Fight Networks with Networks," http://www.rand.org/publications/randreview/issues/rr.12.01/fullalert.html#networks (accessed June 11, 2005). Arquilla and Ronfeldt qualify this: "Al-Qaeda seems to hold advantages at the organizational, doctrinal, and

social levels. The United States and its allies probably hold only marginal advantages at the narrative and technological levels."

8. Clay Shirky, "Power Laws, Weblogs, and Inequality," http://www.shirky.com/writings/powerlaw_weblog.html (accessed June 11, 2005).

9. Gilles Deleuze, *Negotiations*, trans. Martin Joughin (Minneapolis: University of Minnesota Press, 1990), 178.

Nodes

1. John Arquilla and David Ronfeldt, *Networks and Netwars: The Future of Terror, Crime, and Militancy* (Santa Monica: Rand, 2001), 6. A similar litany from 1996 reads: "Netwar is about Hamas more than the PLO, Mexico's Zapatistas more than Cuba's Fidelistas, the Christian Identity Movement more than the Ku Klux Klan, the Asian Triads more than the Sicilian Mafia, and Chicago's Gangsta Disciples more than the Al Capone Gang." See John Arquilla and David Ronfeldt, *The Advent of Netwar* (Santa Monica: Rand, 1996), 5. Arquilla and Ronfeldt coined the term "netwar," which they define as "an emerging mode of conflict (and crime) at societal levels, short of traditional military warfare, in which the protagonists use network forms of organization and related doctrines, strategies, and technologies attuned to the information age." Arquilla and Ronfeldt, *Networks and Netwars*, 6.

2. Mark Wigley, "Network Fever," *Grey Room* 4 (2001).

3. The largest and most important publication series for Internet protocols is called "Request for Comments" (RFC). A few thousand RFC documents have been drafted to date. They are researched, published, and maintained by the Internet Engineering Task Force (IETF) and related organizations.

4. If this section seems overly brief, it is because we have already devoted some attention in other publications to the definition of the concept. See in particular Eugene Thacker, *Biomedia* (Minneapolis: University of Minnesota Press, 2004); and Alexander Galloway, *Protocol* (Cambridge: MIT Press, 2004).

5. Overviews of graph theory are contained in any college-level discrete mathematics textbook. See also Gary Chartrand, *Introductory Graph Theory* (New York: Dover, 1977). For a historical overview, see Norman Biggs et al., *Graph Theory, 1736–1936* (Oxford: Clarendon, 1976). Graph theory principles are commonly used in communications and network routing problems, as well as in urban planning (road and subway systems), industrial engineering (workflow in a factory), molecular biology (proteomics), and Internet search engines.

6. See Paul Baran, *On Distributed Communications* (Santa Monica: Rand, 1964).

7. Gilles Deleuze, *Negotiations* (New York: Columbia University Press, 1995), 178.

8. For instance, in computer culture, specific power relations are articulated by computer users accessing various databases on the Internet. While some of these databases offer public access (e.g., Web-based hubs), others delimit a set of constraints, or differentials in access (e.g., commercial sites such as Amazon, secure Web mail, bibliographic databases at universities, personal e-banking accounts, etc.). Each of these power relations is encompassed by a technology (computers and the Internet), and a force (access to information), and each of them delimits a type of qualitative asymmetry in their power relations (e.g., consumer login to Amazon). From these examples, we see mobilities and constraints, inclusions and exclusions, securities and instabilities. Thus power in this context is less a moral category and more a physico-kinetic category. Power in this sense is less politics and more a kind of physics—a physics of politics.

9. For this reason, the question of substance was a primary concern of medieval philosophy, which sought to explain the relationship between the divine and the earthly, or between spiritual life and creaturely life. While some early thinkers such as Augustine posited a strict distinction between the divine and the earthly, later thinkers such as Aquinas or Duns Scotus were more apt to conceive of a continuum from the lowest to the highest kinds of beings.

10. Aquinas elaborated ten basic kinds of categories in his commentaries on Aristotle and posited an essential link between concept, thing, and word. Later thinkers such as Duns Scotus would complicate this view by suggesting that individuation—at the level of concepts only—proceeded by way of a "contraction" (so that "man" and "animal" are contracted to each other by "rational"), whereas existence as such in the world could not be predicated on anything else.

11. Giorgio Agamben, in writing on the "sovereign exception," cites Walter Benjamin on this point: "The tradition of the oppressed teaches us that the 'state of exception' in which we live is the rule." Giorgio Agamben, *Homo Sacer: Sovereign Power and Bare Life* (Stanford: Stanford University Press, 1998), 55. Agamben takes Benjamin's thesis further by adding that "life is originarily excepted in law" (27). In a sense, Foucault's suggestion that "in order to conduct a concrete analysis of power relations, one would have to abandon the juridical notion of sovereignty" is an affirmation of Agamben's thesis, for Foucault's primary aim is to dismantle an anthropomorphic notion of sovereign power. Michel Foucault, *Ethics: Subjectivity and Truth, vol. 1 of The Essential Works of Michel Foucault* (New York: New Press, 1997), 59.

12. Political thought is remarkably consonant on what constitutes threats to political order—foreign invasion or war is one obvious case, as are disasters

that threaten the political-economic infrastructure of society. But what is striking is how thinkers on opposite sides of the fence politically—such as Hobbes and Spinoza—agree that the greatest threat to political order comes from within: civil war, rebellion, factionalism, and mob rule.

13. Deleuze, *Negotiations*, 180. We note, in passing, that such a networked theory of power is in many ways presaged in Foucault's theses concerning "biopower" in the first volume of *The History of Sexuality*.

14. As the media theorist Vilém Flusser notes, in the network society "we will have to replace the category of 'subject-object' with the category of 'intersubjectivity,' which will invalidate the distinction between science and art: science will emerge as an intersubjective fiction, art as an intersubjective discipline in the search for knowledge; thus science will become a form of art and art a variant of the sciences." Vilém Flusser, "Memories," in *Ars Electronica: Facing the Future*, ed. Timothy Druckrey (Cambridge: MIT Press, 1999), 206.

15. But this is a paradoxical formulation. According to the technical histories of the concept of "information," information cannot matter. Indeed, the familiar associations of cyberspace, e-commerce, virtual identities, and software piracy all have to do with a notion of "information" as disembodied and immaterial, just as the practices of cyberwar and netwar do—and yet with material consequences and costs. Indeed, such a view of information has infused a number of disciplines that have traditionally dealt with the material world exclusively—molecular biology, nanotechnology, immunology, and certain branches of cognitive science.

16. The standards for hardware platforms, operating systems, networking protocols, and database architectures are all examples drawn from the computer and information technology industries. The ongoing development of laboratory techniques, the production and handling of medical data, and policies regarding the distribution and circulation of biological materials are examples in the life sciences.

17. See Lily Kay, *Who Wrote the Book of Life? A History of the Genetic Code* (Stanford: Stanford University Press, 2000); and Evelyn Fox Keller, *Refiguring Life: Metaphors of Twentieth-Century Biology* (New York: Columbia University Press, 1995).

18. David Bourgaize, Thomas Jewell, and Rodolfo Buiser, *Biotechnology: Demystifying the Concepts* (New York: Addison Wesley Longman, 2000), 30.

19. Francis Crick, "On Protein Synthesis," *Symposium of the Society for Experimental Biology* 12 (1958): 144.

20. See Alan Dove, "From Bits to Bases: Computing with DNA," *Nature Biotechnology* 16 (September 1998); and Antonio Regalado, "DNA Computing," *MIT Technology Review*, May–June 2000. Biocomputing includes subareas such as protein computing (using enzymatic reactions), membrane computing (using membrane receptors), and even quantum computing (using quantum fluctuations). Other "nonmedical" applications of biotechnology in-

clude GM foods, chemical synthesis, biomaterials research, biowarfare, and specialized applications in computer science, such as cryptography.

21. See Leonard Adleman, "Molecular Computation of Solutions to Combinatorial Problems," *Science* 266 (November 1994): 1021–24. Also see Adleman's follow-up article "On Constructing a Molecular Computer," *First DIMACS Workshop on DNA Based Computers*, vol. 27 (Princeton: DIMACS, 1997), 1–21. For a more technical review of the field, see Cristian Calude and Gheorghe Paun, *Computing with Cells and Atoms: An Introduction to Quantum, DNA, and Membrane Computing* (London: Taylor and Francis, 2001).

22. The prospect of cellular computing is the most interesting in this respect, for it takes a discipline already working through a diagrammatic logic (biochemistry and the study of cellular metabolism) and encodes a network into a network (Hamiltonian paths onto the citric acid cycle).

23. Compare, for instance, the views of cybernetics, information theory, and systems theory. First, Norbert Wiener's view of cybernetics: "It has long been clear to me that the modern ultra-rapid computing machine was in principle an ideal central nervous system to an apparatus for automatic control." Norbert Wiener, *Cybernetics, or Control and Communication in the Animal and the Machine* (Cambridge: MIT, 1965), 27. Second, Claude Shannon's information theory perspective: "Information must not be confused with meaning. In fact, two messages, one of which is heavily loaded with meaning and the other of which is pure nonsense, can be exactly equivalent, from the present viewpoint, as regards information." Claude Shannon and Warren Weaver, *A Mathematical Theory of Communication* (Chicago: University of Illinois, 1963), 8. Finally, Ludwig von Bertalanffy's biologically inspired systems theory: "The organism is not a static system closed to the outside and always containing the identical components; it is an open system in a quasi-steady state, maintained constant in its mass relations in a continuous change of component material and energies, in which material continually enters from, and leaves into, the outside environment." Ludwig von Bertalanffy, *General Systems Theory: Foundations, Development, Application* (New York: George Braziller, 1976), 121. From the perspective of control, Bertalanffy's work stands in contrast to Wiener's or Shannon's. While von Bertalanffy does have a definition of "information," it plays a much lessened role in the overall regulation of the system than other factors. Information is central to any system, but it is nothing without an overall logic for defining information and using it as a resource for systems management. In other words, the logics for the handling of information are just as important as the idea of information itself.

24. Wiener describes feedback in the following way: "It has long been clear to me that the modern ultra-rapid computing machine was in principle an ideal central nervous system to an apparatus for automatic control.... With the aid of strain gauges or similar agencies to read the performance of

these motor organs and to report, to 'feed back,' to the central control system as an artificial kinesthetic sense, we are already in a position to construct artificial machines of almost any degree of elaborateness of performance." Wiener, *Cybernetics*, 27.

25. As Wiener elaborates, "Just as the amount of information in a system is a measure of its degree of organization, so the entropy of a system is a measure of its degree of disorganization; and the one is simply the negative of the other." Wiener, *Cybernetics*, 11.

26. Shannon and Weaver, *A Mathematical Theory of Communication*, 8.

27. Von Bertalanffy, *General Systems Theory*, 121.

28. Gilbert Simondon, "The Genesis of the Individual," in *Zone 6: Incorporations* (New York: Zone, 1992), 300. In contrast to either atomist (constructionist) or hylomorphic (matter into form) theories of individuation, Simondon's use of the term "individuation" begins and ends with the process of individuation, not its apparent start or end point. Simondon suggests that our electrical technologies of transduction provide a technical means by which material-energetic forms are regulated—individuation is therefore a "transduction."

29. Gilles Deleuze and Félix Guattari, *A Thousand Plateaus* (Minneapolis: University of Minnesota Press, 1987), 6.

30. Gilles Deleuze, *Difference and Repetition*, trans. Paul Patton (New York: Columbia University Press, 1995), 182.

31. Henri Bergson, *The Creative Mind* (New York: Citadel Press, 1997), 147. Another way of stating this is to suggest that networks have no nodes. Brian Massumi corroborates this when he states that "in motion, a body is in an immediate, unfolding relation to its own nonpresent potential to vary. . . . A thing is when it isn't doing." Brian Massumi, *Parables for the Virtual* (Durham: Duke University Press, 2002), 4, 6.

32. John Arquilla and David Ronfeldt, "The Advent of Netwar," in *Networks and Netwars*, 5.

33. Emmanuel Levinas, "Ethics as First Philosophy," in *The Levinas Reader*, ed. Séan Hand (New York: Routledge, 1989), 82–83.

34. Ibid., 83.

35. John Arquilla and David Ronfeldt, *Swarming and the Future of Conflict* (Santa Monica: Rand, 2000), 8.

36. Eric Bonabeau and Guy Théraulaz, "Swarm Smarts," *Scientific American*, March 2000, 72–79.

37. Ibid., 21.

38. Deleuze and Guattari, *A Thousand Plateaus*, 170.

39. Ibid.

40. Foucault offers a distinction between the two types of power near the end of the first volume of *The History of Sexuality*, as well as in a 1976 lecture at the Collège de France: "Unlike discipline, which is addressed to

bodies, the new nondisciplinary power is applied not to man-as-body but to the living man, to man-as-living-being; ultimately, if you like, to man-as-species. . . . After the anatomo-politics of the human body established in the course of the 18th century, we have, at the end of that century, the emergence of something that is no longer an antomo-politics of the human body, but what I would call a 'biopolitics' of the human race." Michel Foucault, *Society Must Be Defended: Lectures at the College de France, 1975–76* (New York: Picador, 2003), 243.

41. For more see Deborah Lupton, *Medicine as Culture: Illness, Disease, and the Body in Western Culture* (London: Sage, 2000); Giorgio Agamben, *Homo Sacer: Sovereign Power and Bare Life* (Palo Alto: Stanford University Press, 1998); Agamben's essay "Form-of-Life," in *Radical Thought in Italy*, ed. Paolo Virno and Michael Hardt (Minneapolis: University of Minnesota Press, 1996); and Michael Hardt and Antonio Negri, *Empire* (Cambridge: Harvard University Press, 2000).

42. See Foucault's texts "The Birth of Biopolitics" and "The Politics of Health in the Eighteenth Century," both in *Ethics: Subjectivity and Truth*, ed. Paul Rabinow (New York: New Press, 1994).

43. As Foucault notes, "After a first seizure of power over the body in an individualizing mode, we have a second seizure of power that is not individualizing, but, if you like, massifying, that is directed not at man-as-body but at man-as-species." Foucault, *Society Must Be Defended*, 243.

44. While theories of media and communication have preferred the term "mass audience" to "population," in the context of the network society, we can see an increasing predilection toward the "living" aspects of networks. Quite obviously, the health care and biomedical research sectors are driven by living forms of all kinds, from "immortalized cell lines" to patients undergoing clinical trials. And despite the rhetoric of disembodied information that characterizes cyberculture, the Internet is still driven by the social and commercial interaction of human beings and "virtual subjects." So while Foucault's use of the term "population" is historically rooted in political economy, we suggest that it is also useful for understanding how the network diagram begins to take shape in a political way. That is because the problem of political economy is also the problem of network management, or what we have called "protocological control."

45. Gilles Deleuze, "What Is a *Dispositif?*" in *Michel Foucault Philosopher* (New York: Routledge, 1992), 164 (translation and emphasis modified).

46. Foucault, "Security, Territory, Population," in *Ethics: Subjectivity and Truth*, 246.

47. Gregor Scott, "Guide for Internet Standards Writers," RFC 2360, BCP 22, June 1998.

48. See Agamben, *Homo Sacer*.

49. See the second chapter of Hardt and Negri, *Empire*.

50. This results in the historical development of a "political science" or a political economy, through which the coordination of resources, peoples, and technologies can be achieved. As Foucault states: "The constitution of political economy depended upon the emergence from among all the various elements of wealth of a new subject: population. The new science called political economy arises out of the perception of new networks of continuous and multiple relations between population, territory and wealth; and this is accompanied by the formation of a type of intervention characteristic of government, namely intervention in the field of economy and population. In other words, the transition which takes place in the eighteenth century from an art of government to a political science, from a regime dominated by structures of sovereignty to one ruled by techniques of government, turns on the theme of population and hence also on the birth of political economy." Foucault, "Governmentality" in *The Foucault Effect: Studies in Govermentality*, ed. Graham Burchell et al. (Chicago: University of Chicago Press, 1993), 100–101.

51. This multistep process is simply a heuristic. To be precise, these steps do *not* happen consecutively. They take place in varying orders at varying times, or sometimes all at once. For example, certain foundational protocols must always precede the genesis of a network (making our step three come before step two). Then after the network is in place, new protocols will emerge.

52. Deleuze, *Negotiations*, 182. The difficulty with relying on Deleuze, however, is that he came to the topic of resisting informatic control rather late in his work (as did Foucault). His work on the topic often includes question marks and hesitations, as if he were still formulating his opinion.

53. Hardt and Negri, *Empire*, 210.

54. Gilles Deleuze, *Foucault* (Minneapolis: University of Minnesota Press, 1988), 92.

55. Ibid.

56. Ibid., translation modified. The quoted phrases refer to Foucault's *History of Sexuality*.

57. In addition, the recurring tropes of AI and "intelligence" (both artificial intelligence and governmental/military intelligence) are made to bolster the dream of a transcendent mind that is not the brain, and a brain that is not the body.

58. D. N. Rodowick, "Memory of Resistance," in *A Deleuzian Century?* ed. Ian Buchanan (Durham: Duke University Press, 1999), 44–45.

59. Political movements oriented around changing existing technologies certainly do exist. We wish not to diminish the importance of such struggles but simply to point out that they are not protocological struggles (even if they are struggles over protocological technologies) and therefore inappropriate to address in the current discussion.

60. Deleuze, *Negotiations*, 175.

61. For a popular overview and discussion of computer viruses, see Stephen Levy, *Artificial Life* (New York: Vintage, 1992), 309.

62. See Fred Cohen, "Computer Viruses: Theory and Experiments," *Computers and Security* 6 (1987): 22–35. Also see Cohen's much-referenced study of computer viruses, *A Short Course on Computer Viruses* (Pittsburgh: ASP Press, 1990).

63. The Web sites of antivirus software makers such as Norton Utilities contain up-to-date statistics on currently operational computer viruses.

64. On computer viruses as a-life, see Eugene Spafford, "Computer Viruses as Artificial Life," in *Artificial Life: An Overview*, ed. Christopher Langton (Cambridge: MIT Press, 2000).

65. These and other SARS figures are contained in the Web sites for the WHO and the CDC. For a recent Rand report on emerging infectious diseases, see Jennifer Brower and Peter Chalk, *The Global Threat of New and Reemerging Infectious Diseases* (Santa Monica: Rand, 2003).

66. See Eugene Thacker, "Biohorror/Biotech," *Paradoxa* 17 (2002); and "The Anxieties of Biopolitics," *Infopeace.org* (Information, Technology, War, and Peace Project) (Winter 2001), http://www.watsoninstitute.org.

67. "It would be neither the fold nor the unfold . . . but something like the *Superfold* [*Surpli*], as borne out by the foldings proper to the chains of the genetic code, and the potential of silicon in third-generation machines. . . . The forces within man enter into a relation with forces from the outside, those of silicon which supersedes carbon, or genetic components which supersede the organism, or agrammaticalities which supersede the signifier. In each case we must study the operations of the superfold, of which the 'double helix' is the best known example." Deleuze, *Foucault*, 131–32.

68. Roland Barthes, *The Pleasure of the Text* (New York: Hill and Wang, 1975), 40.

69. Deleuze goes on to describe how Foucault's work with power reached a certain wall, a limit concerning the silence on the part of those subjected by disciplinary systems such as prisons. This led Foucault to form the GIP (Prisoner's Information Group), opening a new discourse between prisoners, activists, and intellectuals, which decisively informed his work in *Discipline and Punish*. But the same can be said of Deleuze, or anyone doing cultural, social, and political work; one identifies a certain limit point, beyond which something must change. That something could just as easily be concepts as it could be methodology. Or it could be the discarding of a previous set of practices altogether. Further, it could also be a lateral jump from one discipline to another, from a discipline based on theory to one based on practice. Whatever the case, the limit point Deleuze describes is implicit in theoretical work, and this is our responsibility here in addressing protocological control.

70. Deleuze, "Intellectuals and Power," 205–6.

71. Geert Lovink, *Dark Fiber* (Cambridge: MIT Press, 2002), 9.

Edges

1. Michael Fortun, "Care of the Data," a talk given at the symposium *Lively Capital 2: Techno-corporate Critique and Ethnographic Method*, University of California, Irvine, October 23–24, 2005.

2. Michel Foucault, *The Hermeneutics of the Subject: Lectures at the Collège de France, 1981–82* (New York: Palgrave Macmillan, 2005), p. 11.

3. Alan Liu, *The Laws of Cool* (Chicago: University of Chicago Press, 2004), 331.

4. Ibid.

5. Plato, *Republic*, trans. Allan Bloom (New York: Basic Books, 1991), 222 (8.544c–d).

6. Ibid., 243 (8.564b–c).

7. Michel Foucault, *"Society Must Be Defended"* (New York: Picador, 2003), 34.

8. Ibid., 39.

9. Ibid., 50.

10. Jean-Luc Nancy, *The Birth to Presence* (Stanford: Stanford University Press, 1993), 36.

11. Ibid., 39.

12. Albert-László Barabási, *Linked* (Cambridge: Perseus, 2002), 221.

13. Ibid.

14. James Beniger, *The Control Revolution* (Cambridge: Harvard University Press, 1989), 247.

15. See Philip Agre, "Surveillance and Capture: Two Models of Privacy," *Information Society* 10, no. 2 (1994): 101–27.

16. Gilles Deleuze and Félix Guattari, *A Thousand Plateaus* (Minneapolis: University of Minnesota Press, 1987), 29.

17. The "fork" command is a function that allows a piece of software to spawn a clone of itself. If called recursively, each clone will also clone itself, thereby creating a "fork bomb" within the computer's memory. If left unchecked, a fork bomb will crash its host computer extremely quickly. The fork bombs displayed here, however, autoprotect themselves from crashing; they strangle the machine but do not kill it. Written in Perl, a powerful text-parsing language created in the late 1980s by Larry Wall, these fork bombs output generative textures in ASCII text, then quit and return the machine to normal functionality. By creating a high-stress environment within the computer's processor, the artifacts of the machine itself become visible in the output. These scripts are inspired by the work of Alex McLean, Jaromil, and Jodi.

18. Foucault, *"Society Must Be Defended,"* 243–44.

19. Martin Heidegger, *Being and Time*, trans. Joan Stambaugh (Albany: SUNY Press, 1996), 45.

20. Ibid., 46.

21. Barabási, *Linked*, 16.

22. On December 1, 2003—World AIDS Day—Health and Human Services officials directly linked diseases and war. Just as troops in Iraq were "saving people from tyranny," so were U.S. health agencies "saving people from disease."

23. The CDC has had a number of network-based programs under way that address this network-response challenge. The Enhanced Surveillance Project and the National Electronic Disease Surveillance System are two examples.

24. Beniger, *The Control Revolution*, 20.

25. Critical Art Ensemble, *The Electronic Disturbance* (New York: Autonomedia, 1994), 12.

26. Julian Stallabrass, "Just Gaming: Allegory and Economy in Computer Games," *New Left Review* 198 (March–April 1993): 104.

27. Vilém Flusser, *Writings* (Minneapolis: University of Minnesota Press, 2002), 20.

28. Geert Lovink, *My First Recession* (Rotterdam: V2, 2003), 14.

29. Ibid., 47.

30. Karl Marx, *Economic and Philosophic Manuscripts of 1844*, trans. Martin Milligan (New York: Prometheus, 1988), 75–76. Marx also speaks of the instruments of labor in a way that recalls his earlier formulation of the "inorganic body," in effect placing the inorganic body of the 1844 manuscripts within the context of capitalism: "Leaving out of consideration such ready-made means of subsistence as fruits, in gathering which a man's bodily organs alone serve as the instruments of his labor, the object the worker directly takes possession of is not the object of labour but its instrument. Thus nature becomes one of the organs of his activity, which he annexes to his own bodily organs." Karl Marx, *Capital, Volume I*, trans. Ben Fowkes (1867; New York: Penguin, 1990), 285.

31. Friedrich Nietzsche, *The Will to Power*, trans. Walter Kaufmann and R. J. Hollingdale (New York: Vintage, 1968), 659.

32. Michel Foucault, *The Birth of the Clinic* (New York: Vintage, 1973), 153.

33. Walter Benjamin, *Illuminations* (New York: Schocken Books, 1968), 239.

34. Emmanuel Levinas, *On Escape* (Stanford: Stanford University Press, 2003), 55.

35. Ibid., 54.

36. Ibid., 67.

37. Agamben, *The Coming Community*, 86.

38. Paul Virilio, *The Aesthetics of Disappearance* (New York: Semiotext(e), 1991), 101.

39. Ibid., 58.

40. Hakim Bey, *TAZ: The Temporary Autonomous Zone* (Brooklyn: Autonomedia, 1991), 132.

41. The "extradimensional biological" stories of Lovecraft, Clark Ashton Smith, and Frank Belknap Long stand out in this subgenre. The other

references in this paragraph are to *Faust* (dir. Jan Svankmajer, 1994); *The Thing* (dir. John Carpenter, 1982); H. P. Lovecraft, "The Dunwich Horror" and "The Call of Cthulhu," in *The Call of Cthulhu and Other Weird Stories* (New York: Penguin, 1999); and *Begotten* (dir. Elias Merhinge, 1990).

42. Aristotle, *De anima*, trans. Hugh Lawson-Tancred (New York: Penguin, 1986), 2.2, p. 159.

43. Thomas Aquinas, *Summa theologiae*, 1.22.2, in *Selected Writings*, ed. and trans. Ralph McInerny (New York: Penguin, 1998).

44. Henri Bergson, "The Perception of Change," in *The Creative Mind* (New York: Citadel, 1974), 147.

45. Raoul Vaneigem, *A Declaration of the Rights of Human Beings* (London: Pluto, 2004), 31–32.

46. Lyotard, *The Inhuman: Reflections on Time* (Cambridge: Polity, 1991), 77.

47. Deleuze and Guattari, *A Thousand Plateaus*, 411.

48. Luis Villareal, "Are Viruses Alive?" *Scientific American*, December 2004, 105.

49. Roland Barthes, *Image–Music–Text* (New York: Hill and Wang, 1977), 149.

50. See Georges Bataille, *The Accursed Share*, vol. 1 (New York: Zone, 1998).

Coda

1. Hardt and Negri, *Multitude*, 222.

2. Ibid., 99.

3. In this characterization, Hobbes would represent the side of secular sovereignty (in fact, sovereignty is a precondition for the commonwealth to exist at all), whereas Spinoza would represent the side of the multitude (following on his ontology, which emphasizes radical immanence, "God or nature," as the basic principle of all reality). But this is, of course, only a heuristic opposition. Hobbes, for all his railing against the "multitude not yet united into one person," shows a great deal of ambivalence about the role that the multitude plays (this is especially evident in *De cive*). Sometimes it is the threat of instability within the commonwealth (the "disease" of civil war), and sometimes the multitude is necessary for the passage from the "state of nature" to a fully formed commonwealth. Likewise, while contemporary readings of Spinoza often radicalize him as a proponent of the multitude, texts such as the *Tractatus Theologico-Politicus* show an equally ambivalent attitude toward the multitude: sometimes it is a revolutionary, almost self-organizing force, and at other times it is simply factionalism and mob rule.

4. Paolo Virno, *A Grammar of the Multitude*, trans. Isabella Bertoletti (New York: Semiotext(e), 2004), 25.

5. John Arquilla and David Ronfeldt, "The Advent of Netwar (Revisited)," in *Networks and Netwars* (Santa Monica: Rand, 2001), 6.

6. Ibid., 2.

7. Ibid., 20.

8. "Afterword (September 2001): The Sharpening Fight for the Future," in *Networks and Netwars*, 314.

9. Ibid., 317.

10. In the earlier text, they write that empire is "a dynamic and flexible systemic structure that is articulated horizontally," and in the later text they describe the multitude in similar language: "The global cycle of struggles develops in the form of a distributed network. Each local struggle functions as a node that communicates with all the other nodes without any hub or center of intelligence." Hardt and Negri, *Empire*, 13; and Hardt and Negri, *Multitude*, 217. The following from *Empire* is also indicative: "In contrast to imperialism, Empire establishes no territorial center of power and does not rely on fixed boundaries or barriers. It is a *decentered* and *deterritorializing* apparatus of rule that progressively incorporates the entire global realm within its open, expanding frontiers. Empire manages hybrid identities, flexible hierarchies, and plural exchanges through modulating networks of command" (xii–xiii). In *Multitude* they write that "the new global cycle of struggles is a mobilization of the common that takes the form of an open, distributed network, in which no center exerts control and all nodes express themselves freely" (218). The minichapter on the White Overalls ends with strong confirmation of the design, shape, and characteristics of the distributed network form: "What may have been most valuable in the experience of the White Overalls was that they managed to create a form of expression for the new forms of labor—their networked organization, their spatial mobility, and temporal flexibility—and organize them as a coherent political force against the new global system of power" (267). As we have noted, flexibility and increased mobility are both important qualities of distributed networks. Further, they describe how the "magic of Seattle" was realized in a "network structure. The network defines both their singularity and their commonality. . . . The various groups involved in the protests [are linked] in an enormous open network" (288).

11. "The fact that a movement is organized as a network or swarm does not guarantee that it is peaceful or democratic." Hardt and Negri, *Multitude*, 93.

12. Ibid., 68, 87. In fact, Hardt and Negri suggest that there must be some sort of formal harmony between the two historical actors, writing that resistance is to have the "same form" as the dominant and that the two should "correspond."

13. Hardt and Negri recognize this as "a sort of abyss, a strategic unknown" in their own work: "All notions that pose the power of resistance as homologous to even similar to the power that oppresses us are of no more use" (*Multitude*, 90). This is what terrorism has done to U.S. foreign policy, driving the government to revolutionize rapidly from a model of neoliberalism and engagement to a model of global networked sovereignty. The suicide bomber does

not achieve slow, deliberate reform of military occupation. Rather, it revolutionizes the entire landscape of occupation, generally evolving the level of conflict higher and devolving the occupying army toward unconventional guerrilla tactics. But these are horrifying examples. The Internet was also an asymmetrical threat: in the 1960s, Paul Baran revolutionized the entire nature of how communications switching could happen in a network (and hence the historical shift from decentralized to distributed communications switching).

14. This is, in many ways, one of the primary challenges of thinking about networks—if networks are in some fundamental sense "unhuman," then this means that any attempt to think about networks will confront the horizon of thinking itself.

Appendix

1. Friedrich Kittler, *Gramophone, Film, Typewriter* (Stanford: Stanford University Press, 1999), 3 (italics added).

2. This has been achieved already by the Beige programming ensemble's "Netbuster" product.

Electronic Mediations

Katherine Hayles, Mark Poster, and Samuel Weber, Series Editors

Index

distributed network. *See* network: distributed

distribution, 17, 31, 34, 64, 66, 68–69, 78, 94, 97, 111, 179n13. *See also* network: distributed

distribution, power law. *See* power law distribution

disturbance, 101

diversity, 84. *See also* heterogeneity

dividual, 39–40. *See also* individual; individuation

DNA, 6, 28, 48–53, 54, 77, 79, 90, 122–23, 132–33

DNA chip, 50

DNA computer, 51–53, 58

Domain Name System (DNS), 44–45, 47, 54

Dope Wars, 115

double helix, 175n67

doubt, 160

Dr. Faustus, 138

drift, 163

Duns Scotus, John, 169n9, 169n10

dystopia, 134

Economic and Philosophical Manuscripts of 1844 (Marx), 177n30

economy: Fordist, 3; post-Fordist, 2, 107; service, 3

edge, 22, 31–32, 37, 39, 52, 59–60, 62, 99, 112, 139, 142–43, 146, 157

Edwards, Sean, 67

élan vital, 80, 101

elemental, 155–57

e-mail, 10, 28, 83, 95, 145–46, 169n8

emergence, 22, 30, 47, 69

emerging infectious disease. *See* disease: emerging infectious

emp, 163

empire, 6, 8, 27, 152, 179n10

Empire (Hardt and Negri), 152

emptiness, 142

empty, 160

endemic, 115–17

end-to-end principle, 47

enemy. *See* enmity

Enhanced Surveillance Project, 177n23

enmity, 63–70, 94, 154

entropy, 172n25

envision, 164

Enzensberger, Hans Magnus, 16

epidemic, 75, 86, 105–9, 115–17, 129

epidemiC, 105

epidemiology, 88, 92, 94

equality, 13

error, 129

essence, 134–35

Ethernet, 42

exception, 38–39, 101, 162; sovereign, 169n11; state of, 110, 169n11. *See also* topology: exceptional

exceptionalism, 22, 39, 149; American, 3, 14

excess, 147

existence, 136–38

exorcise, 164

exploit, 17, 20–22, 46, 81–97, 135, 145–46, 155, 157

exploitation, 78–79, 135, 159, 166

face, 12, 64–70. *See also* defacement; facing; interface

faciality, 68

facing, 64–70. *See also* defacement; face; interface

Fahrenheit 911, 2

fail, 164

failure, 96–97, 155, 157

family, 22

FedEx, 9

feedback, 55, 122–24, 171n24

feminism, 14

Fidelista, 26, 168n1

TCP/IP. *See* Internet protocol;
    transmission control protocol
technology, 3, 10, 26–27, 31–35,
    51, 56, 70, 75–76, 81, 84, 98–99,
    112, 125–26, 131, 135–37, 139,
    143, 167n7
telecommunications, 55
television, 123–24
TELNET, 42
temporary autonomous zone,
    137–38
terrorism, 9, 11, 13, 15–17, 27, 115–
    18, 179n13. *See also* bioterror-
    ism; terrorist; war on terror
terrorist, 20, 40, 64, 112, 134, 151.
    See also cyberterrorist; network;
    terrorism
Thales, 156
Them!, 69
theology, 131
thing, 37
Time, 90
topology, 13–14, 22, 32, 34, 39,
    59, 61, 63–64, 65–70, 88, 94,
    97, 112, 138, 143, 151–52,
    156, 161; exceptional, 22, 40,
    44–45, 149
Toronto, 4, 89–91
totality, 29, 154–55
TP53 gene, 129
Tractatus Theologico-Politicus
    (Spinoza), 178n3
transgression, 91, 95, 97
transmission, 130
transmission control protocol
    (TCP), 29, 42–43, 45, 54, 96,
    123, 125, 144
transport layer, 42–43
trash, 145–47
traveling salesman problem, 51–52
Trojan horse, 83
Tron, 69
trust, 45
Tzara, Tristan, 147

unhuman. *See* nonhuman
unilateralism, 7–9
United Nations, 2, 6–8, 11, 116
United States, 6–8, 11, 14–15, 20–
    21, 59, 89–91, 113, 116, 120–21,
    154, 167n7
universal, 9–10, 22–23, 29, 74, 98,
    131–32, 160
universe, 144
unknown, 133–34, 161, 179n13
unmanned aerial vehicle, 15, 41
unordered, 162
user, 143
U.S. Department of Health and
    Human Services, 120, 177n22
U.S. Department of Homeland
    Security, 8, 16–17, 154
U.S. military, 16, 25
U.S. Navy, 84
U.S. News, 90
utopia, 134

vaccine, 119–20
Vancouver, 90
Vaneigem, Raoul, 139
vector, 97, 161
vectoralist, 135
Venice Biennale, 105
vertex. *See* node
verticality, 25–26, 39
Vietnam, 89–90, 151
violence, 18
viral aesthetics. *See* aesthetics, viral
Virilio, Paul, 9, 137–38
Virno, Paolo, 150
virus, 119–22, 142; biological 4, 83,
    86–87, 105–9, 118; computer, 4,
    12, 30, 75, 82–85, 87, 94–96, 99,
    105–9, 119–22, 154, 175n63. *See
    also* worm, Internet
vitalize, 162
von Bertalanffy, Ludwig, 56–57,
    171n23
von Clausewitz, Carl, 64, 98, 111

Alexander R. Galloway is assistant professor in the Department of Media, Culture, and Communication at New York University. He is the author of *Gaming: Essays on Algorithmic Culture* (Minnesota, 2006) and *Protocol: How Control Exists after Decentralization*.

Eugene Thacker is associate professor of new media in the School of Literature, Communication, and Culture at the Georgia Institute of Technology. He is the author of *Biomedia* (Minnesota, 2004) and *The Global Genome: Biotechnology, Politics, and Culture*.